# DICTIONARY OF
# SCIENTIFIC
# WORD ELEMENTS

# DICTIONARY OF
# SCIENTIFIC
# WORD ELEMENTS

*Chemistry—Mathematics—Physics*

WILLIAM B. MULLEN

Georgia Institute of Technology

1969

## LITTLEFIELD, ADAMS & CO.
Totowa, New Jersey

GEORGE E. MAYCOCK

2951

# Preface

I should like to thank those whose help has made this book possible: my wife, Eleanor, for her invaluable assistance and encouragement; and the Georgia Institute of Technology for awarding me funds from the National Science Foundation Institutional Grant for Science, so that I could complete this work.

William B. Mullen
*Atlanta, Georgia*

# Introduction

The purpose of this dictionary is to reduce the vocabulary burden of science and engineering students by demonstrating that a large number of technical and scientific terms are built up from a small number of basic word elements. As the student learns to analyze scientific terms into their components and becomes familiar with these basic word elements, he will find that he can more easily learn and remember the technical terms of his basic science and mathematics courses.

Since the word elements listed in this dictionary are also components of many non-technical or general vocabulary words, the student will find that his study of these scientific word elements will lead to increased proficiency in the area of his general vocabulary. In order to suggest the wider usefulness of these scientific word elements, a selection of two thousand general vocabulary words derived from these scientific word elements is included in this dictionary.

This dictionary has been arranged for most efficient use by the student. In the main body of the dictionary will be found an alphabetical listing of the principal word elements making up the specialized vocabularies of chemistry, mathematics, and physics. Each word element is illustrated by at least two scientific terms, which are analyzed and defined, and a selection of related general vocabulary words. At the end of the main section appear two short tables of General Scientific Suffixes and Chemical Suffixes, and an Index of all scientific terms defined or illustrated in the main section.

Under each word element heading in the main section appear at least two representative terms with the appropriate science or sciences indicated by (C) chemistry, (M) mathematics, or (P) physics; a few non-specific terms from general science are indicated by (S). Following each scientific term are listed in square brackets all significant word elements, and below each term is a concise definition. Multiple definitions are listed in the order of the sciences indicated in parentheses and are separated by semicolons. If the definition does not make clear the relationship between the word elements and the term, an explanation is provided in parentheses. At the end of each word element section appears in parentheses a selection of general vocabulary words derived from the given word element.

The method of displaying the word elements composing a scientific term requires fuller explanation. Ordinarily the component word elements (one hyphen indicating a prefix element and two indicating a root element) are simply placed within square brackets and joined by a plus sign. Omission of part of an element is indicated by parentheses. Meaningless connective vowels between word elements are omitted:

SONIC (P)  [-SON-(sound)]

TRIGONOMETRY  (M)  [-TRI-(three)  +  -GON-(angle)
  + -METR-(measure)]

AMBIENT  (C, P)  [AMB(I)-(around)  +  -IENT-(go)]

Quotation marks indicate that an English word rather than a word element has been used in the formation of the term:

ANTICATHODE  (P)  [ANTI-(opposite)  +  "cathode"]

AUTOXIDATION  (C)  [-AUT(O)-(self)  +  "oxidation"]

Capital letters within quotation marks or within foreign words indicate that only the capitalized section appears in the scientific term:

EXTRAPOLATE  (M)    [EXTRA-(beyond)  +
    "interPOLATE"]

AG (C)  [Latin: ArGentum(silver)]

PERPENDICULAR  (M)  [Latin: perpendiculum(plumb
    line)  <  PER-(an intensive)  +  PENDere(hang)]

The arrowhead symbol <, meaning "derived from," is used

to indicate a spelling modification of a word element or the word element source of a quoted word or foreign language word. The first two examples below illustrate the common process of assimilation, where the final letter of a prefix blends with the initial letter of the following word element:

IMMISCIBLE (C)   [IM < IN-(not) + -MISC-(mix)]

IRREGULAR (M)   IR < IN-(not) + "regular" < -REGUL-(rule)]

INHIBITOR (C)   [IN-(in) + HIB < -HAB-(hold)]

COROLLARY (M)   [Latin: corolla(little crown, garland) < CORONa(crown)]

A colon followed by "thus" is used to bridge the gap between the literal meaning of the word elements and the definition of the term:

ELIMINATE (M)   [E-(out) + -LIM-(threshold: thus put out of doors, expel)]

OBTUSE (M)   [OB-(against) + -TUS-(beat: thus blunted)]

DERIVATION (M)   [DE-(from) + -RIV-(stream: thus drawing off water, obtaining from a source)]

IMPEDANCE (P)   [IM < IN-(in) + -PED-(foot: thus literally entangling the feet, and thus hindering)]

Foreign language words are listed only when necessary to explain an unusual spelling or a meaning which differs from the word element meaning:

HG (C)   [Latin: HydrarGyrum(mercury) < -HYDR-(water, liquid) + -ARG-(silver)]

CALCULATE (M)   [Latin: calculus(pebble, used as a counter in doing arithmetic) < CALCis(limestone)]

MOMENTUM (P)   [Latin: momentum(movement) < movimentum < -MOV-(move)]

GRAM (C, P)   [Latin: gramma(a small weight, from the marking thereon) < Greek: GRAMma(writing)]

MEAN (M)   [French: meien < Latin: medianus < MEDius(middle)]

## A- (not, without)

ADIABATIC (P)  [A- + DIA-(through) + -BAT-(pass)]
    of changes in volume or pressure occurring without gain or loss of heat

AMORPHOUS (C)  [A- + -MORPH-(form)]
    without definite shape or crystalline structure

ANEROID (P)  [A- + -NER-(wet)]
    of a barometer not using liquid mercury

ANHYDROUS (C)  [AN- + -HYDR-(water)]
    of a compound without water or water of crystallization in its composition

ASTATINE (C)  [A- + -STAT-(stand, rest)]
    element (so named because it is unstable)

ASYMPTOTE (M)  [A- + SYM-(together) + -PTOT-(fall)]
    straight line which continually approaches but never meets a curve

*(abyss, agnostic, amnesty, amoral, anarchy, anonymous, apathy, atheism)*

## AB- (from, away)

ABERRATION (P)  [AB- + -ERR-(wander)]
    failure of a lens or mirror to bring all light rays to a single focus

ABSCISSA (M)  [AB- + -SCISS-(cut)]
    horizontal distance of a point from the Y-axis (so named because literally a segment cut off from the X-axis); compare ORDINATE

ABSORPTION (C, P)  [AB- + -SORPT-(suck in)]
    process of taking up and retaining internally a liquid or gas by a porous substance; compare ADSORPTION; process of taking in and transforming radiant energy into a different form.

1

ABSTRACT (S)  [AB(S)- + -TRACT-(pull: thus separated
    or apart from the practical or concrete)]
    pure or theoretical; opposed to APPLIED

*(abduct, abhorrent, abnormal, aborigines, abrupt, absent,
    abstinence)*


### -AC- (sharp)

ACID (C)  [-AC-(sharp: thus sharp to the taste, sour)]
    compound that is able to react with a base to form a salt
    (so named because in water solution it tastes sour)
ACUTE (M)  [-AC-]
    of an angle less than 90°; compare OBTUSE

*(acerbity, acme, acrid, acrimony, acrobat, acumen, acuity,
    exacerbate)*


### -ACT- (*see* -AG-)


### -ACTIN- (ray, radiation)

ACTINISM (C)  [-ACTIN-]
    property of radiant energy that produces chemical changes
ACTINIUM (C)  [-ACTIN-]
    radioactive element

*(actinoid, actinolite, actinotherapy)*


### AD- (to)
### (also AC-, AP-, AT-, etc., depending on following letter)

ACCELERATOR (C, P)  [AC < AD- + -CELER-(hasten)]
    substance which speeds up a chemical reaction; compare
    CATALYST; device for increasing the velocity of charged
    particles to break apart atoms

ADHESION (P)  [AD- + -HES-(stick)]
   force holding together molecules of dissimilar substances
   in contact; compare COHESION
ADJACENT (M)  [AD-(near to) + -JAC-(lie)]
   having a common side and vertex, as ADJACENT ANGLES
ADSORPTION (C)  [AD- + -SORPT-(suck up)]
   taking up a substance at the solid surface of another; compare ABSORPTION
APPLIED (S)  [AP < AD- + PLY < -PLIC-(fold: thus
   attached to)]
   put to practical use; opposed to ABSTRACT
ATTRACTION (P)  [AT < AD- + -TRACT-(draw, pull)]
   mutual force between bodies causing them to approach
   each other or resist separation; opposed to REPULSION

   *(accede, adequate, adherent, adverb, adverse, affiliate,
      annihilate, aspire)*


### -AERO- (air, gas)

AERODYNAMICS (P)  [-AERO- + "dynamics" <
   -DYN-(power)]
   branch of physics dealing with the laws of motion of air
   and other gases and the forces exerted on bodies by such
   motion
AEROSOL (C)  [-AERO- + "SOLution" < -SOLUT-
   (loosen)]
   dispersion of solid or liquid particles in a gas

   *(aerate, aerial, aerolite, aeronautics, aerospace, airily,
      aria, debonair)*


### -AG-, -ACT- (do, drive)

COAGULATION (C)  [CO-(together) + -AG-(drive)]
   formation of a solid or soft, gelatinous mass from a liquid
   by chemical reaction

3

REACTANCE (P)  [RE-(back) + -ACT-(do)]
   opposition by capacitance and induction of a circuit to the
   flow of alternating current
REACTION (C, P)  [RE-(back) + -ACT-(do)]
   change involving action of substances upon each other;
   force acting in opposition to a given force
REAGENT (C)  [RE-(back) + -AG-(do)]
   substance used to measure or change another by means of
   their mutual chemical action

*(agile, agitate, cogent, counteract, exact, exigent, pedagogue, transact)*


## -ALB- (white)

ALBEDO (P)  [-ALB-]
   percentage of total light reflecting from a planet or satel-
   lite
ALBUMEN (C)  [-ALB-]
   protein found in white of egg; also class of simple proteins

*(alb, albatross, albescent, albino, Albion, auburn, daub)*


## -ALLO-, -ALI- (other)

ALIQUOT (C, M)  [-ALI- + -QUOT-(how many)]
   measured proportion of the volume of a solution; dividing
   into another number without remainder; opposed to ALI-
   QUANT
ALLOMERISM (C)  [-ALLO- + -MER-(part)]
   variation in chemical composition without difference in
   crystalline form
ALLOTROPY (C)  [-ALLO- + -TROP-(change)]
   property of an element existing in different forms with
   different properties

4

PARALLEL (M, P) [PAR(A)-(beside) + -ALLEL-(one another) < -ALLO-]
of lines or planes extending in same direction and equidistant at all points; of an electrical hookup in which like poles or terminals are connected

*(alias, alibi, alien, allegory, allergy, allonym, allopathy, allophone)*


## AMBI-, AMPHO- (both)


AMBIENT (C, P) [AMB(I)-(on both sides, thus around)+ -IENT-(going)]
of the surrounding medium, as AMBIENT TEMPERATURE
AMPHOTERIC (C) [AMPHO-(both)]
exhibiting both basic and acidic properties

*(ambiguity, ambidextrous, ambition, ambivalence, amphibian, amphitheater)*


## -AMPLI- (large)


AMPLIFICATION (P) [-AMPLI- + -FIC-(make)]
increase in voltage or power of an electronic signal
AMPLITUDE (P) [-AMPLI-]
maximum extent of an electrical wave or oscillation measured from the mean

*(ample, ampliation, amplify, AM radio)*


## ANA- (up, according to)


ANALOG (M) [ANA-(according to) + -LOG-(proportion)]
class of computers using numbers represented by measur-

5

able physical quantities; compare DIGITAL COMPUTER using numbers expressed directly as digits

ANALYSIS (C, M)   [ANA-(up) + -LYS-(loosen)]
determining nature or proportion of constituent parts of a substance by separating the ingredients; working out of problems by means of equations or examining the relations of variables

ANODE (C, P)   [AN(A)-(up) + -OD-(way)]
positive electrode, through which current enters a non-metallic conductor; opposed to CATHODE ("down way"); also electronic tube electrode collecting electrons

ANODIC (C, P)   ["anode"]
of an element above hydrogen in the electrochemical series (thus tending to form an anode); opposed to CATHODIC

*(anachronism, anabolism, anagram, analogous, anathema, anatomy, aneurysm)*

## ANTI- (opposite, against)

ANTICATHODE (P)   [ANTI- + "cathode"]
target anode in an electron tube which is opposite the cathode

ANTILOGARITHM (M)   [ANTI- + "logarithm"]
number which corresponds to a given logarithm

ANTIOXIDANT (C)   [ANTI- + "oxidant"]
substance having the property of opposing oxidation

ANTIPARTICLE (P)   [ANTI- + "particle"]
mirror image or counterpart of an ordinary particle of matter, as ANTINEUTRON, ANTIPROTON, POSITRON

*(antagonism, anticlimax, antipathy, antiseptic, antinomy, antithesis)*

APO- (C)   [APO-]
>   prefix denoting a compound derived from or related to, as APOMORPHINE

APOCHROMATIC (P)   [APO- + -CHROM-(color)]
>   of a lens that is completely free from chromatic aberration (signifying a higher degree of correction than ACHROMATIC)

APOGEE (P)   [APO- + -GE-(earth)]
>   greatest distance from earth of orbiting heavenly body or vehicle; opposed to PERIGEE

APOTHEM (M)   [APO- + -THEM-(placed)]
>   perpendicular from center to one side of a regular polygon

*(apocalypse, apostasy, apostle, apostrophe, apotheosis, apothecary)*


## -AQUA- (water)

AQUA REGIA (C)   [-AQUA- + -REG-(royal)]
>   mixture of nitric and hydrochloric acids (so named because it dissolves the noble metals, gold and platinum)

AQUEOUS (C)   [-AQU(A)-]
>   of a solution using water as a solvent

*(aquacade, aquamarine, aquarium, aquatic, aqueduct, eau de, ewer, sewage)*


## -ARG- (silver)

AG (C)   [Latin: ArGentum(silver)]
>   symbol for the element silver

ARGENTIC (C)   [-ARG-]
>   containing silver in its highest valence; compare ARGENTOUS

7

HG (C)   [Latin: HydrarGyrum(mercury) < -HYDR-(water, liquid) + -ARG-]
    symbol for the element mercury
LITHARGE (C)   [-LITH-(stone) + -ARG-]
    lead monoxide (so named because obtained as a by-product of smelting silver-bearing ores)

*(argent, argentine, Argentina, argil, argillaceous, Argyrol)*


## -ARITHM- (number)

ARITHMETIC (M)   [-ARITHM-]
    science or art of computing with numbers, involving mainly addition, subtraction, division, multiplication
LOGARITHM (M)   [-LOG-(proportion) + -ARITHM-]
    power to which a fixed number, or base, must be raised to produce a given number

*(none)*


## -ASTR(O)- (star, heavens)

ASTROID (M)   [-ASTR-]
    hypocycloid of four cusps that resembles a star
ASTRONAUTICS (P)   [-ASTRO- + -NAUT-(sailor)]
    science of operating space vehicles
ASTRONOMY (P)   [-ASTRO- + -NOM-(law)]
    science of the heavenly bodies: size, history, motion, constitution, etc.
ASTROPHYSICS (P)   [-ASTRO- + "physics" < -PHYS-(nature)]
    branch of astronomy dealing with physical and chemical constitution and properties of the heavenly bodies

*(aster, asterisk, asterism, asteroid, astral, astrolabe, astrology, disaster)*

8

## -AUDI- (hear)

AUDIBILITY (P)   [-AUDI-]
   ratio of the strengths of a transmitted signal and a barely audible signal, expressed in decibels

AUDIO- (P)   [-AUDI-]
   combining form indicating hearing, sound, as AUDIOGRAM, AUDIOMETER

*(audience, audio-visual, audit, audition, auditorium, obeisance, obey, oyez)*

## -AUR- (gold)

AU (C)   [Latin: AUrum(gold)]
   symbol for the element gold

AURIC (C)   [-AUR-]
   of compounds containing gold in its highest valence; compare AUROUS

*(aureate, aureole, Aureomycin, El Dorado, oriole, ormolu, orphrey)*

## -AUTO- (self, self-acting)

AUTOCATALYSIS (C)   [-AUTO- + "catalysis" < CATA- (an intensive) + -LYS-(loosen)]
   increase in rate of reaction caused by one of its products

AUTOCLAVE (C)   [-AUTO- + -CLAV-(key)]
   gastight vessel for high-pressure reactions (so named originally from its literal sense of self-fastening)

AUTOMATION (S)   ["automatic" < -AUTO- + -MAT-(moving)]
   use of machines to control other machines

9

AUTOXIDATION (C)   [-AUT(O)- + "oxidation"]
   oxidation that occurs on exposure to air

   *(autarky, autocracy, autograph, autonomous, automat, automaton, autopsy)*


## -BAR- (pressure, weight)


BAR (P)   [-BAR-(pressure)]
   cgs international unit of pressure
BARIUM (C)   [-BAR-(weight)]
   element (so named because not occurring free but in the alkaline earth BARYTA, characterized by its great weight)
BAROMETER (P)   [-BAR-(pressure) + -METER-(measure)]
   instrument for measuring atmospheric pressure
ISOBAR (P)   [-ISO-(equal) + -BAR-(weight, pressure)]
   any of two or more atoms having the same atomic weight but different atomic numbers; line connecting points of equal barometric pressure

   *(baritone, barysphere)*


## -BAS- (foundation)


BASE (C, M)   [-BAS-]
   in chemistry, substance which can enter into combination with an acid to neutralize it and form a salt; in geometry, assumed line or plane on which a figure rests; in mathematics, a number raised to a power or a constant figure on which a mathematical table depends, as in logarithms or decimal system

   *(basal metabolism, baseless, basement, basic, basis)*

10

## -BAT- (pass)

ADIABATIC (P)  [A-(not) + DIA-(through) + -BAT-]
   of changes in volume or pressure effected without gain or
   loss of heat
ANABATIC (P) [ANA-(up) + -BAT-]
   of an upward-moving current of warm air; opposed to
   KATABATIC (KATA-"down")

*(acrobat, aerobatics, diabetes)*


## BI- (two, double)

BI- (C)  [BI-]
   prefix indicating double proportion, as BICARBONATE; or, in
   organic compounds, the doubling of a radical, as BIPHENYL
   (also DI-)
BIMETAL (P)  [BI- + "METALlic"]
   device using bond of two dissimilar metals
BINARY (C, M)  [Latin: BINARius(double) < BI-]
   of a compound having two different elements or radicals;
   of a system of numbers using base 2
BINOMIAL (M)  [BI- + -NOMI-(name)]
   algebraic expression consisting of two terms connected by
   a plus or minus sign

*(biennial, bifocal, bigamy, bilateral, bilingual, binocular,
   combine)*


## -BI(O)- (life)

BIO- (C, P)  [-BIO-]
   combining form indicating concern with living organisms,
   as BIOASTRONAUTICS, BIOCHEMISTRY, BIOPHYSICS

BIONICS (P)   [-BI(O)- + "electrONICS"]
    study of living systems as models for the development of
new computer and electronic systems

*(amphibious, anaerobic, antibiotic, biography, biopsy,*
*microbe, symbiosis)*


## -BOL- (throw)


BOLOMETER (P)   [-BOL-(throw: thus light beam, radia-
    tion) + -METER-(measure)]
    instrument for measuring very small amounts of radiant
energy
HYPERBOLA (M)   [HYPER-(beyond) + -BOL-(throw:
    thus a throwing beyond, excess)]
    curve formed by the intersection of a plane with a right
circular cone (so named because the inclination of the
plane to the base of the cone exceeds that of the side of
the cone)
PARABOLA (M)   [PARA-(beside) + -BOL-]
    curve formed by the intersection of a cone with a plane
parallel to a side of the cone (so named because its axis is
parallel to side of cone)
SYMBOL (C, M, P)   [SYM-(together) + -BOL-(throw:
    thus sense of compare)]
    conventional sign standing for a quantity, operation, ele-
ment, etc.

*(ballista, ballistic, diabolic, embolism, metabolism, parable,*
*problem)*


## -BROM- (stink)


BROM(O)- (C)   ["BROMine"]
    combining form indicating bromine as a principal element
in compounds, as BROMOFORM, BROMOIODIDE

12

BROMINE (C)   [-BROM-]
    element (so named for its extremely unpleasant odor)

    *(bromide, bromidic, bromidrosis, bromomania)*


### -CALC- (limestone)


CALCINE (C)   [-CALC-]
    drive off volatile matter by heating to a high temperature
    without fusing (so named from the process of reducing
    limestone to quicklime)

CALCIUM (C)   [-CALC-]
    element (so named because it never occurs free but in
    such combinations as chalk, limestone, etc.)

CALCULATE (M)   [Latin: calculus(little stone, pebble,
    used as a counter in doing arithmetic) < CALCis
    (limestone)]
    compute, determine by arithmetical means

CALCULUS (M)   [Latin: calculus(little stone, pebble)]
    method of calculation using a specialized system of
    algebra

    *(calcify, calcimine, calcography, chalk, incalculable)*


### -CALOR- (heat)


CALORESCENCE (P)   [-CALOR-]
    incandescence produced by the absorption of heat radia-
    tions

CALORIE (C)   [-CALOR-]
    unit of heat in the cgs system

CALORIFIC (VALUE) (C)   [-CALOR- + -FIC-(make)]
    quantity of heat produced by complete combustion of a
    unit weight of a fuel

CALORIMETER (C)   [-CALOR- + -METER-(measure)]
  instrument for measuring amounts of absorbed or evolved
  heat

  *(cauldron, calefactory, calenture, caloric, nonchalant, scald)*


### -CAND- (glow)

CANDLE (P)   [-CAND-]
  unit of luminous intensity
INCANDESCENCE (P)   [IN-(in) + -CAND-]
  emission of light because of high temperature

  *(candelabrum, candid, candidate, candor, chandelier,*
  *incendiary, incense)*


### -CAPILL- (hair)

CAPILLARITY (P)   [-CAPILL-]
  form of surface tension between liquid molecules and
  solid molecules (so named because observable in a capillary tube)
CAPILLARY (P)   [-CAPILL-]
  of a tube with a very small (hairlike) bore

  *(capillaceous, capilliculture, capilliform, depilatory)*


### -CAP(T)-, -CEPT- (take)

CAPACITANCE (P)   [-CAP-(take: thus able to hold)]
  property of circuit or body which permits the storage of
  an electrical charge
CAPTURE (P)   [-CAPT-]
  process by which an atom or nucleus receives an additional particle
INTERCEPT (M)   [INTER-(between) + -CEPT-]
  include between two points, lines, or planes

14

SUSCEPTIBILITY (P)   [SUS < SUB-(under) + -CEPT-
(take: thus to receive)]
ratio of the magnetization of a substance to the strength
of the magnetizing force, thus a measure of the capacity
of a substance for being magnetized

*(accept, capable, capacious, captivate, conception, except,
receptacle)*

-CARBO- (carbon)

CARBOHYDRATES (C)   [-CARBO- + -HYDR-(water)]
group of organic compounds containing carbon combined
with hydrogen and oxygen in the proportion 2:1
CARBON (C)   [Latin: CARBONis(coal)]
element

*(carbonaceous, carbonado, carbonation, carboniferous,
carbuncle, carburetor)*

CAT(A)- (down)

CATALYSIS (C)   [CATA-(down, thus "completely") +
-LYS-(loosen)]
speeding up of reaction by adding substance (CATALYST)
which remains unchanged
CATHODE (C, P)   [CAT(A)- + -HOD-(way)]
negative electrode, through which current leaves a non-
metallic conductor; opposed to ANODE ("up way"); also
electronic tube electrode emitting electrons
CATHODIC (C)   ["cathode"]
of an element below hydrogen in the electrochemical
series (thus tending to form a cathode); opposed to
ANODIC

15

KATABATIC (P)   [KATA<CATA- + -BAT-(pass)]
of a downward-moving wind cooled by radiation; opposed
to ANABATIC ( ANA-"up")

*(cataclysm, catalepsy, catalogue, cataract, catastrophe,
catholic, katabasis)*


### -CATEN- (chain)

CATENARY (M)   [-CATEN-]
curve assumed by a cord or chain hanging from two fixed
points
CATENATION (C)   [-CATEN-]
linkage between atoms of the same element

*(catena, catenulate, concatenation)*


### -CAUST- (burn)

CAUSTIC (C)   [-CAUST-]
capable of burning or eating away human tissue by chemical action; note also CAUSTIC POTASH, CAUSTIC SODA
CAUSTIC (SURFACE) (P)   [-CAUST-]
surface formed by the ultimate intersection of rays emitted
from one point and reflected or refracted from a curved
surface (so named because the intensity of the light, and
thus the heat, is generally greater at a point on this surface, and at certain points could initiate combustion)

*(caustically, cauterize, encaustic, holocaust, ink)*


### -CAV- (hollow)

CAVITATION (P)   [-CAV-]
formation of low-pressure vapor cavities in a moving fluid

CONCAVE (P)   [CON-(an intensive) + -CAV-]
  of a lens which is hollow and curved, as the inside of a
  sphere; opposed to CONVEX

  *(cage, cave, cavern, cavernous, cavity, decoy, excavation)*


## -CED-, -CESS- (go)

ANTECEDENT (M)   [ANTE-(before) + -CED-]
  first term of a ratio

PRECESSION (P)   [PRE-(before) + -CESS-]
  slow gyration of the rotation axis of a spinning body, as
  a gyroscope or top, due to action of a torque tending to
  change the direction of the rotation axis (so named from
  the analogous motion of the earth's axis which causes the
  precession, or earlier occurrence, of the equinoxes in each
  sidereal year)

  *(access, ancestor, excessive, intercede, proceed, recession,
  secede)*


## -CELER- (hasten)

ACCELERATION (P)   [AC < AD-(to) + -CELER-]
  rate of change in velocity per unit of time

ACCELERATION (OF GRAVITY) (P)   [AC < AD-(to)+
  -CELER-]
  increase in velocity of a freely falling body due to force
  of gravity

ACCELERATOR (C, P)   [AC < AD-(to) + -CELER-]
  substance which speeds up a chemical reaction, compare
  CATALYST; device for increasing the velocity of charged
  particles to break apart atoms

17

DECELERATION (P)  [DE-(reversal) +
   "acCELERATION"]
   negative acceleration, rate of decrease of velocity per unit
   of time

*(accelerative, celerity)*


## -CELES- (sky)

CELESTIAL (MECHANICS) (P)  [-CELES-]
   branch of astronomy concerned with the forces of gravitation as they affect the motions of celestial bodies
CELESTITE (C)  [-CELES-]
   the mineral strontium sulfate (so named from its blue
   color)

*(ceiling, celesta, Celeste, Celestial City, Celestial Empire,
cerulean)*


## -CENT- (one hundred)

CENTI- (C, P)  [-CENT-]
   prefix in metric system meaning 1/100, as CENTIGRAM,
   CENTIMETER, etc.
CENTIGRADE (C, P)  [-CENT- + -GRAD-(step)]
   of a temperature scale having 100 degrees between freezing and boiling point of water

*(cent, centenary, centennial, centipede, centurion, century,
percentage)*


## -CENTR- (center)

CENTRIFUGAL (FORCE) (P)  [-CENTR- + -FUG-(flee)]
   inertial reaction tending to cause bodies to move away
   from a center it revolves about; compare CENTRIPETAL
   FORCE

18

CONCENTRATE (C)   [CON-(together) + -CENTR-]
    increase the relative proportion of desired material in a
    solution or mixture by the removal of undesired material

CONCENTRIC (M)   [CON-(together) + -CENTR-]
    having same center; opposed to ECCENTRIC

ECCENTRIC (M)   [EK < EX-(out) + -CENTR-]
    not having the same center, as circles; opposed to CON-
    CENTRIC

*(anthropocentric, central, centrifuge, concentration,
    decentralize, eccentric)*

**-CEPT-** (*see* **-CAP(T)-**)

**-CESS-** (*see* **-CED-**)

**-CHEMI-, -CHEMO-**  (chemical)

CHEMISORPTION (C)   [-CHEMI- + "adSORPTION" <
    AD-(to) + -SORPT-(suck up)]
    adsorption which depends on chemical and not physical
    attraction

CHEMOSPHERE (P)   [-CHEMO- + "atmoSPHERE" <
    -ATMOS-(vapor) + -SPHER-(ball)]
    stratum of the atmosphere characterized by photochemi-
    cal activity; compare STRATOSPHERE

*(chemotherapy)*

**-CHLOR-**  (green, chlorine)

CHLORINE (C)   [-CHLOR-]
    element (so named from its green color); note use in
    compounds: CHLORIC, CHLOROUS, CHLORATE

19

CHLOROPHYLL (C)   [-CHLOR- + -PHYLL-(leaf)]
green coloring matter present in the leaves of plants

*(chlorosis, chlorella, chlorinate, chloroform, chlorophane)*

## -CHOR- (volume)

ISOCHOR (P)   [-ISO-(equal) + -CHOR-]
plotted curve showing the relationship between pressure and temperature of a substance at constant volume

PARACHOR (C)   [PARA-(along with) + -CHOR-]
function expressing constant relationship between surface tension and density of a substance, and which is proportional to the molecular volume

*(chorography)*

## -CHROM- (color)

CHROMA (P)   [-CHROM-]
purity or intensity of a color

CHROMATIC (ABERRATION) (P)   [-CHROM-]
lens defect causing the colors in a beam of light to be focused at different points; compare ACHROMATIC (A- "without"), of a lens transmitting or refracting light without separating it into its component colors

CHROMIUM (C)   [-CHROM-]
element (so named because of its brightly colored compounds); note also use in compounds: CHROMATE, CHROMIC

DICHROISM (C, P)   [DI-(two) + -CHRO(M)-]
property of solutions being differently colored in different concentrations; property of crystals showing different colors from different directions

*(achromic, chromatic, chromo, chromosome, monochrome,
polychrome)*

20

## -CHRON- (time)

BRACHISTOCHRONIC (P)  [-BRACHISTO-(shortest) + -CHRON-]
of the minimum time path for a falling body between two points in space

SYNCHRONOUS (P)  [SYN-(together) + -CHRON-]
having the same period and phase

SYNCHROTRON (P)  [SYN-(together) + -CHRO(N)- + -TRON(device)]
apparatus for accelerating atomic particles (so named because the accelerating forces synchronize with the movement of the particles)

TAUTOCHRONE (P)  [-TAUTO-(the same) + -CHRON-]
inverted cycloid representing force of gravity: time of descent from every point to lowest point is the same

*(anachronism, chronic, chronicle, chronology, chronometer, synchronize)*

## -CID- (fall)

COINCIDE (M)  [CO-(together) + -CID-]
have the same dimension and place in space

INCIDENCE (M, P)  [IN-(on) + -CID-]
partial coincidence between two figures; falling of a light ray or projectile on a surface, or the angle of falling

*(accident, cadence, coincidental, decadence, deciduous, incident, occident)*

## CIRCUM- (around)

CIRCUIT (P)  [CIRCU(M)- + -IT-(go)]
complete electric or magnetic path

CIRCUMSCRIBE (M)  [CIRCUM- + -SCRIB-(write, draw)]
   draw a figure around another with maximum number of contact points

*(circumlocution, circumscribe, circumspect, circumstance, circumvent)*


## -CLIN- (lean, slope)

DECLINATION (P)  [DE-(from) + -CLIN-]
   angular distance of a heavenly body measured north or south from the celestial equator
INCLINATION (M)  [IN-(on) + -CLIN-]
   angle measured between two intersecting lines or planes; in analytic geometry, the angle made by a line measured counterclockwise from the positive X-axis

*(clinic, enclitic, inclined, decline, recline)*


## -CLUD-, -CLUS- (shut)

EXCLUSION (PRINCIPLE) (P)  [EX-(out) + -CLUS-]
   principle that no two electrons can occupy the same orbit, i.e., have the same set of quantum numbers
INCLUDED (M)  [IN-(in) + -CLUD-]
   of the angle between two straight lines

*(cloister, conclude, disclose, exclusive, preclude, recluse, seclusion)*


## CO(M)- (together; (M)complement of)
(also COL-, CON-, COR-, etc., depending on following letter)

COEFFICIENT (M, P)  [CO- + EF< EX-(out) + -FIC- (make: thus work out together, cooperate)] number or symbol placed before and multiplying an algebraic expression; number expressing kind and quantity of change in a process or substance under specified conditions

COFUNCTION (M) [CO-(complement of) + "function"] corresponding trigonometric function of the complementary angle, as COSINE, COTANGENT, etc.

COLLIMATE (P) [COL < COM- + LIM < -LIN-(line)] make rays of light parallel; adjust line of sight of an optical instrument

COMPLEMENT (M) [COM-(together, thus "completely") + -PLE-(fill)]
amount of angle or arc which will complete a right angle or 90°

CONSTANT (C, M, P) [CON < COM-(together, thus "completely") + -STA-(stand)]
property or quality that remains the same under the same conditions, or quantity that remains the same throughout the calculations

CORRELATION (M) [COR < COM- + "relation"] degree of relationship between variables or sets of data

*(colloquial, commiserate, compassion, consonant, conspire, cooperate)*

### -COLL- (glue-like)

COLLODION (C) [-COLL- + OD < -EID-(appearance)] solution of nitrated cellulose in ether and alcohol, used as a coating for wounds and as a membrane for dialysis

COLLOID (C) [-COLL-]
state of matter in which minute particles of a solid are suspended in a liquid

*(collotype, protocol)*

## -COM- (hair)

COMA (P)  [-COM-]
> hazy border surrounding point of light or image of object, caused by spherical aberration in a lens

COMET (P)  [Greek: kometes(long haired: thus describing comet's tail)]
> heavenly body orbiting the sun, consisting of a relatively condensed nucleus and a long, tenuous tail

*(none)*


## CONTRA- (*see* COUNTER-)


## -CORON- (crown)

COROLLARY (M)  [Latin: corolla(little crown, garland)< CORONa(crown)]
> proposition following from another without new proof (so named from sense of gift or gratuity, which developed from original sense of a present of a garland)

CORONA (P)  [-CORON-]
> luminous discharge surrounding a conductor under high voltage; ring of light surrounding sun or moon

*(corolla, Corona Borealis, coronary, coronation, coroner, coronet)*


## -COSM- (universe)

COSMIC (RAYS) (P) [-COSM-]
> high energy particles bombarding the earth from outer space

COSMOGONY (P)  [-COSM- + GON < -GIN-(be born)]
> study of the origin and formation of the universe

COSMOTRON (P)  [-COSM- < "COSMic rays" + -TRON
     (device)]
   high energy accelerator of charged particles comparable
   to cosmic rays
MICROCOSMIC (SALT) (C)  [-MICRO-(small) +
     -COSM-]
   white, crystalline salt used as a reagent in blowpipe
   analysis (so named because originally obtained from the
   urine of man, considered as a microcosm or miniature of
   the world)

   *(cosmetic, cosmic, cosmos, cosmonaut, cosmopolitan,
            cosmopolite, macrocosm)*


## COUNTER-, CONTRA- (opposite, against)

CONTRAST (P)  [CONTRA- + -ST(A)-(stand)]
   brightness, color, or sound intensity difference
COUNTER (EMF) (P)  [COUNTER-]
   electromotive force developed in some circuits which
   opposes the normal flow of current

   *(contradict, contrary, contravene, controversy, counteract,
            counterpart)*


## -CREM- (grow)

DECREMENT (M)  [DE-(from, away) + -CREM-]
   decrease in value of a variable
INCREMENT (M)  [IN-(an intensive) + -CREM-]
   amount by which a variable increases between two successive variables

   *(accretion, concretize, crescent, crescendo, excrescence,
            increscent)*

## -CREP- (crackle)

DECREPITATE (C)  [DE-(an intensive) + -CREP-]
   heat salts and minerals until crackling sound is caused or
   ceases

DISCREPANCY (S)  [DIS-(apart) + -CREP-(crackle: thus
       differ in sound, disagree)]
   lack of agreement in parallel procedures or operations

*(crepitation, crevice, decrepit, decrepitude, discrepant)*


## -CRU- (cross)

CRUCIBLE (C)  [-CRU-]
   vessel made of heat-resistant substances for melting metal
   or minerals (so named from original sense of light burning
   before a church cross, hence a hanging lamp, and hence
   an earthen pot)

CRUNODE (M)  [-CRU- + -NOD-(loop)]
   point where a curve crosses itself and thus has two tan-
   gents

*(crux, crucial, crucifix, crucify, cruise, crusade, excruciating)*


## -CRY(O)- (freezing)

CRYOGENICS (P)  [-CRYO- + -GEN-(produce)]
   branch of physics concerned with extremely low temperatures

CRYOSCOPY (C)  [-CRYO- + -SCOP-(observe)]
   determination of properties of substances through study
   of freezing points of liquids

*(cryolite, cryotherapy, crystal)*

# -CRYPT- (hidden)

CRYPTOCRYSTALLINE (C)   [-CRYPT- + "crystalline" < -CRY(O)-(frost)]
   having a crystalline structure too small to be seen under a microscope; opposed to PHANEROCRYSTALLINE
KRYPTON (C)   [Greek: krypton(hidden)]
   element (so named because its presence in minute quantities in the atmosphere was undetected for so long)

*(apocryphal, crypt, cryptic, cryptogram, cryptography, cryptonym, grotto)*

# -CUPR- (copper)

CU (C)   [Latin: CUprum(copper) < aes cyprium(metal of Cyprus)]
   symbol for the element copper; note use in compounds CUPRIC, CUPROUS
CUPRO- (C)   [-CUPR-]
   combining form meaning copper, as CUPROMAGNESITE, CUPRONICKEL

*(cupreous, cupriferous)*

# -CURR- (run)

CONCURRENT (M)   [CON-(together) + -CURR-]
   of lines meeting at a point
CURRENT (P)   [-CURR-]
   flow of a fluid, or flow of electricity in a conductor; compare AC, DC

*(cursory, discursive, excursion, incur, occur, precursor, recur, succor)*

27

## -CURV- (bent)

CURVATURE (M)   [-CURV-]
   rate of deviation of an arc from a straight line tangent to
   it
CURVILINEAR (M)   [-CURV- + -LIN-(line)]
   made up of or enclosed by curved lines

   *(curb, curvaceous, curvet, incurvate)*

## -CYBERN-, -GOVERN- (control)

CYBERNETICS (P)   [Greek: kybernetes(helmsman, pilot)]
   comparative study of the principles of control and communication in the human nervous system and electronic
   control devices
GOVERNOR (P)   [-GOVERN-]
   device for automatically controlling the speed of an engine
   or motor

   *(governance, gubernatorial)*

## -CYCL- (circle)

CYCLE (P)   [-CYCL-]
   one complete series of a periodic process such as current
   alternation, oscillation, etc.
CYCLIC (C)   [-CYCL-]
   characterized by a ring or closed chain structure of atoms
CYCLOID (M)   [-CYCL-]
   curve traced by a point on a circle rolling along a straight
   line; compare EPICYCLOID, HYPOCYCLOID
CYCLOTRON (P)   [-CYCL- + -TRON(device)]
   apparatus for accelerating atomic particles (so named because of the circular movement of the particles)

   *(bicycle, cyclist, cyclone, Cyclops, cyclorama, encyclical,*
   *encyclopedia)*

## -DAT-, -DON- (give)

DATUM (M)   [-DAT-]
  magnitude, figure, or relation given or known from which other magnitudes, figures, or relations can be deduced or calculated

DONOR (C)   [-DON-]
  substance which can give up part of itself for combination with another substance

  *(antidote, data, date, dative, donation, extradite, pardon, tradition)*

## DE- (from, reversal)

DECANTATION (C)   [DE-(from) + -CANT-(lip of a cup)]
  separation of a liquid from sediment or higher density liquid by gently pouring off the supernatant liquid

DECELERATION (P)   [DE-(reversal) + "acCELERATION"]
  negative acceleration, decrease in velocity per unit time

DEPENDENT (VARIABLE) (M)   [DE-(from) + -PEND-(hang: thus relying on)]
  variable whose value is a function of another, called INDEPENDENT VARIABLE

DEVIATION (M, P)   [DE-(from) + -VIA-(road)]
  difference between the mean of a series of data and an individual member; bending of radiation from a straight path

  *(decapitate, décolleté, deduction, defame, degenerate, detraction, deviant)*

## -DEC- (ten)

DECA- (S)   [-DEC-]
  combining form in metric system meaning 10, as DECA-

29

LITER, DECAMETER; also in geometry, as DECAGON, DECA-
HEDRON

DECI- (S)   [-DEC-]
combining form in metric system meaning 1/10, as
DECIGRAM, DECILITER; also note DECIBEL

DECILE (M)   [-DEC-]
any value in a series which divides the distribution of the
individuals in the series into ten groups of equal frequency, or any of the groups

DECIMAL (M)   [-DEC-]
based on the number 10, or a fraction whose unwritten
denominator is 10 or a power of 10

*(decade, Decalogue, Decameron, decathlon, December,
decimate, decussate)*

### -DENS- (crowded)

CONDENSATION (C, P)   [CON-(together) + -DENS-]
reaction involving union between atoms of same or different molecules to form a new, more complex compound;
changing from gaseous to liquid state or liquid to solid
state

CONDENSER (P)   [CON-(together) + -DENS-]
system of lenses for concentrating light rays; device for
temporarily storing electricity, a capacitor

DENSIMETER (P)   [-DENS- + -METER-(measure)]
instrument for measuring specific gravity of liquids

DENSITY (P)   [-DENS-]
mass of a substance per unit volume; quantity of electric
current per unit area flowing in a conductor

*(dense, condense, contrail)*

## -DEUTER- (second)

DEUTERIUM (C)   [-DEUTER-]
   isotope of hydrogen with atomic weight of approximately
   two
DEUTERON (C)   [-DEUTER-]
   nucleus of a deuterium atom

*(deuteragonist, deuterocanonical, deuterogamy, deuterogenesis,
Deuteronomy)*

## -DEXTR- (right)

DEXTROROTATORY (C)   [-DEXTR- + "rotatory" <
   -ROTA-(turn)]
   of a substance which turns the plane of polarization of
   light to the right; opposed to LEVOROTATORY
DEXTROSE (C)   [-DEXTR- + "glucOSE"]
   dextrorotatory form of glucose, also called grape sugar or
   corn sugar

*(ambidextrous, dexterity)*

## DI- (two)

DI- (C)   [DI-]
   prefix indicating two atoms, molecules, radicals, etc., as
   DIOXIDE, DICHLORIDE
DIPOLE (P)   [DI- + "pole" < -POL-(axis)]
   system consisting of equal and opposite electrical charges
   or magnetic poles

*(dichotomy, diglot, dilemma, diphthong, diploma, diptych,
dissyllable)*

31

## DIA- (through)

DIAGONAL (M)  [DIA- + -GON-(angle)]
  straight line joining two nonadjacent angles of a figure

DIAMETER (M, P)  [DIA- + -METER-(measure)]
  straight line which passes through the center of a circle or
  sphere from one side to the other; unit of measurement of
  the magnifying power of a lens (equal to the number of
  times the linear size of the object is increased)

DIATHERMANCY (P)  [DIA- + -THERM-(heat)]
  property of being capable of transmitting infrared or heat
  rays

DIELECTRIC (P)  [DI(A)- + "electric"]
  material permitting passage of lines of force of an electric
  field but not conducting current

*(diabetes, diagnose, dialect, dialogue, diametrical, diaphanous,
diaphragm)*

## -DIDYM- (twin)

DIDYMIUM (C)  [-DIDYM-]
  mixture of rare earth elements (including praesodymium
  and neodymium), formerly considered one of the elements, used to color glass for optical filters (so named
  because of its close association or twin brotherhood with
  lanthanum)

PRASEODYMIUM (C)  [-PRASEO-(green) +
  "diDYMIUM"]
  element (so named because salts of this component of
  didymium are usually green); the other component element was named NEODYMIUM (NEO-"new")

*(none)*

# DIS- (apart)

DIFFRACTION (P)  [DIF < DIS- + -FRACT-(break)]
  breaking up and scattering of light rays by partial obstruction or in passing by edges of opaque bodies

DISINTEGRATION (P)  [DIS-(apart, thus reversal) + -INTEG-(whole)]
  change in the structure of an atomic nucleus, as through emission of nucleons by a radioactive element

DISPERSION (C, M, P)  [DI(S)- + -SPERS-(scatter)]
  scattering and suspension of fine particles (DISPERSED PHASE) in a liquid medium (DISPERSION MEDIUM); scattering from their average of a series of values in a frequency distribution; separation of light into different colors by refraction

DISSOCIATION (C)  [DIS- + -SOCI-(join)]
  breakdown of a compound into simpler constituents, or of an electrolyte into ions

  *(diffident, discord, discrete, disgrace, disruption, dissect, distend)*

## -DON- (*see* -DAT-)

## -DUC(T)- (to lead)

CONDUCTION (P)  [CON-(together) + -DUCT-]
  transmission of heat, sound, or electricity through matter by the passage of energy from particle to particle; note also CONDUCTANCE, ability of a substance to transmit electricity without loss of heat or light energy

INDUCTION (P)  [IN-(in) + -DUCT-]
  production of electrical or magnetic effects in a conductor or magnetizable body in proximity to the influence or variation of a field of force

33

REDUCTION (C, M) [RE-(back) + -DUCT-]
originally, changing a substance to another, usually simpler form, such as bringing an ore to the metallic state or removing oxygen from a compound; by extension, decreasing the positive valence or increasing the negative valence of an element or radical; in mathematics, changing an expression to a simpler form

TRANSDUCER (P) [TRANS-(across) + -DUC-]
device supplying power in a different form to a second system

*(abduction, adduce, conducive, deduce, educe, introduce,*
*produce, seduce)*

## -DULC- (sweet)

DULCIFY (C) [-DULC- + FY < -FIC-(make)]
make free from acidity or saltness, sweeten

EDULCORATE (C, M) [E-(out) + -DULC-(sweet: thus to sweeten by removing)]
eliminate soluble acids or other impurities by washing; in computer technology, to remove worthless information

*(dulcet, dulciana, dulcimer, dulcinea)*

## -DUR- (hard)

DURALUMIN (C) [-DUR- + "ALUMINum"]
alloy of aluminum and copper (so named because comparable in hardness and strength to soft steel)

DUROMETER (C) [-DUR- + -METER-(measure)]
instrument for measuring the hardness of metals

*(durable, duress, endure, indurate, obdurate)*

## -DYN- (power)

DYNAMICS (P)   [-DYN-]
   branch of physics dealing with effects of forces on bodies
   in motion or at rest; compare STATICS
DYNE (P)   [-DYN-]
   unit of force in the cgs system
HETERODYNE (P)   [-HETERO-(different) + -DYN-]
   combine radio waves of different frequencies to produce
   beat frequencies
THERMODYNAMICS (P)   [-THERM-(heat) + -DYN-]
   branch of physics dealing with the relationship between
   heat and other forms of energy

   *(dynamic, dynamism, dynamite, dynamo, dynasty)*

## -DYS- (difficult)

DYSOXIDIZE (C)   [-DYS- + "oxidize"]
   oxidize with difficulty
DYSPROSIUM (C)   [-DYS- + -PROS-(approach)]
   element (so named because found in minute amounts in
   minerals together with other rare earths)

*(dysentery, dyspeptic, dysphoria, dysteleology, dysthanasia)*

## E- (*see* EX-)

## -ELAST- (springy)

ELASTICITY (P)   [-ELAST-]
   tendency of a body to resist deforming forces and to
   return to its original shape

35

ELASTOMER (C)   [-ELAST- + "polyMER" < -POLY-
(many) + -MER-(part)]
class of polymers with elastic, rubberlike properties, as the
synthetic rubbers

*(elastic, elasticized, elastration, elastosis, inelastic)*

## -ELECTR- (electric)

DIELECTRIC (P)   [DI(A)-(through) + "electric" < Latin:
electrum(amber, or the electric charge produced in it
by friction)]
material permitting passage of lines of force of an electric
field but not conducting current
ELECTRET (P)   [-ELECTR- + "magnET"]
solid with a permanent electric polarization
ELECTRODE (P)   [-ELECTR- + -OD-(way)]
conducting element in an electron tube, transistor, battery,
arc lamp
ELECTROLYTE (C)   [-ELECTR- + -LYT-(dissolve)]
substance which in solution dissociates into free ions and
conducts electricity

*(electrify, electrocute, electronic)*

## EPI- (upon)

EPICYCLOID (M)   [EPI- + -CYCL-(circle)]
curve traced by a point on a circle rolling on the outside
of a fixed circle; compare HYPOCYCLOID
PARALLELEPIPED (M)   [-PARALLEL-(parallel) + EPI-
+ -PED-(ground, base)]
prism whose bases are parallelograms

*(epidermis, epigram, epilepsy, Episcopal, epitaph, epithet,
epitome)*

EQUATION (C, M)   [-EQU-]
> statement of equality between reacting substances and products; statement of equality between quantities or expressions

EQUI- (M)   [-EQUI-]
> combining form meaning equal, as EQUIANGULAR, EQUIDISTANT, EQUILATERAL

EQUILIBRIUM (C, P)   [-EQUI- + -LIBR-(balance)]
> state of even balancing between rates of forward and reverse reactions, or between forces whose resultant is zero

EQUIVALENT (C, M)   [-EQUI- + -VAL-(strength)]
> having same valence or combining weight; having equal area but of different shape

*(equable, equanimity, equate, equinox, equitable, equity, equivocal)*


## -ERG- (work)

ARGON (C)   [A-(not) + -ERG-(work: thus idle, inert)]
> inert gaseous element

CHEMURGY (C)   ["CHEMistry" + URG < -ERG-]
> branch of applied chemistry dealing with industrial development of new products from organic raw materials such as farm products

ENERGY (P)   [EN-(in) + -ERG-]
> capacity for doing work and overcoming inertia

ERG (P)   [-ERG-]
> unit of work and energy in the cgs system

*(allergy, energetic, energize, exergue, organ, synergetic, thaumaturgy)*

## -ERR- (wander)

ABERRATION (P)   [AB-(away) + -ERR-]
  departure from an ideal light ray path, causing imperfect optical image

ERROR (M)   [-ERR-]
  difference between observed or calculated value of a magnitude and the true value

  *(aberrant, err, errant, errata, erratic, erroneous)*

## -ERT-, -ART- (skill)

INERT (C)   [IN-(not) + -ERT-(skill: thus unskilled, and also inactive, lazy)
  having little or no chemical action, ability to combine

INERTIA (P)   [IN-(not) + -ERT-]
  tendency of a body to persist in a state of rest or motion

*(artifact, artifice, artificer, artificial, artisan, artless, inertness)*

## EU- (good)

EUDIOMETER (C)   [EU- + -DIA-(weather) + -METER-(measure)]
  instrument for determining volume and composition of gases (so named because originally tested quantity of oxygen in air, supposedly higher in fair weather)

EUTECTIC (C)   [EU- + -TECT-(melt)]
  mixture which has lowest possible melting point

*(eugenics, eulogy, eupeptic, euphemism, euphonious, euphoria, euthanasia)*

38

# EX(O)-, E- (out)

ECCENTRIC (M)   [EK < EX- + -CENTR-(center)]
    not having the same center, as circles; opposed to CON-
    CENTRIC

EMISSION (P)   [E-(out) + -MISS-(send)]
    giving off of energy or charged particles; note also the
    acronym LASER, from the initial letters of "Light Amplifi-
    cation by Stimulated Emission of Radiation"

EXOTHERMIC (C)   [EXO- + -THERM-(heat)]
    of a change accompanied by or produced from heat libera-
    tion; opposed to ENDOTHERMIC

EXPONENT (M)   [EX- + -PON-(place: thus literally placed
    out from a quantity)
    number or symbol placed above and to the right of an
    expression to signify an operation to be performed

*(ebullient, eccentric, emissary, exclude, exorbitant, exponent,*
*extant)*

## EXTRA- (beyond)

EXTRAPOLATE (M)   [EXTRA- + "interPOLATE"]
    project beyond the range of known values; compare
    INTERPOLATE

EXTREME (M)   [Latin: extremus < exterus(outside) >
    EXTRA-]
    initial or final term of a proportion or series

*(ESP, extracurricular, extraneous, extravagant, extrinsic,*
*extroversion)*

## -FACE- (surface)

FACET (C)   [-FACE-]
    any small plane surface of a crystal

INTERFACE (P)   [INTER-(between) + -FACE-]
surface forming boundary between two adjacent parts of matter or space

*(deface, efface, facing, facade, superficial)*

### -FAC(T)- (*see* -FIC-)

### -FER- (strike)

INTERFERENCE (P)   [INTER-(between, among) + -FER-]
mutual action of waves of sound, light, etc. in augmenting or neutralizing each other

INTERFEROMETER (P)   ["INTERFERence" + -METER-(measure)]
instrument using interference phenomena of light to measure wavelengths or to analyze sections of a spectrum

*(ferule, interfere)*

### -FER-, -LAT- (carry)

ABLATION (P)   [AB-(away) + -LAT-]
melting or vaporization of expendable parts to carry away excess heat from essential parts, as nose cone materials during reentry

CIRCUMFERENCE (M)   [CIRCUM-(around) + -FER-]
boundary line of a circle or curvilinear figure

OBLATE (M)   [OB-(opposite; in opposite direction from "prolate") + -LAT-(carry: thus extended at the equator)]
flattened at the poles; opposed to PROLATE

PROLATE (M)   [PRO-(forward) + -LAT-]
extended at the poles; opposed to OBLATE

*(conference, differ, infer, offer, refer, related, superlative, translate)*

## -FERR- (iron)

FE (C)   [Latin: FErrum (iron)]
   symbol for the element iron; note use in compounds
   FERRIC, FERROUS
FERRO- (C)   [-FERR-]
   combining form indicating of or alloyed with iron as
   FERROMANGANESE, or containing ferrous iron as FERRO-
   CYANIDE

   *(farrier, ferrotype, ferruginous, ferrule)*

## -FERV- (boil)

EFFERVESCENT (C)   [EF < EX-(out) + -FERV-]
   giving off bubbles of gas
FERMENTATION (C)   [FERM < -FERV-]
   process of gradual decomposition of organic compounds
   through the action of various ferments

   *(fervid, fervent, fervor, perfervid)*

## -FIC-, -FAC(T)- (make, do)

COEFFICIENT (M, P)   [CO-(together) + EF < EX-(out)
      + -FIC-(make: thus to work out together, cooperate)]
   number or symbol placed before and multiplying an
   algebraic expression; number expressing kind and quantity
   of change in a process or substance under specified conditions
EFFICIENCY (P)   [EF < EX-(out) + -FIC-(make: thus
      to work out, accomplish)]
   ratio of work done or energy expended to energy supplied
   in producing it, output over input
FACTOR (M)   [-FACT-]
   one of two or more quantities which, multiplied together,
   form a product

41

RAREFACTION (C, P)   [-RAR-(scarce) + -FACT-(make)]
   process of making gases less dense; region of minimum
   pressure in a compression wave medium

   *(artificial, defect, efficacy, facsimile, proficient, significant,
   suffice)*


## -FIGUR- (form)

CONFIGURATION (P)   [CON-(together) + -FIGUR-
   (form: thus pattern)]
   spatial arrangement of atoms in a molecule or particles in
   an atom
FIGURE (M)   [-FIGUR-]
   surface or space enclosed by lines or planes

   *(disfigure, effigy, figment, figurative, figurine, prefigure,
   transfigure)*


## -FIL- (thread)

FILAMENT (P)   [-FIL-]
   slender wire in a light bulb or electron tube
FILAR (P)   [-FIL-]
   of instruments such as micrometer or microscope having
   fine threads or hairs across the field of view

   *(bifilar, defilade, enfilade, filigree, fillet, profile, purfle)*


## -FIN- (limit)

AFFINITY (C)   [AF < AD-(to) + -FIN-(limit: thus bor-
   dering upon, related)]
   selective attraction between differing chemical elements
   or groups of elements which brings about formation of
   new compounds

42

DEFINITION  (P)   [DE-(from)  +  -FIN-(boundary: thus
set a limit to, show precise outlines of)]
power of a lens to give fine detail, a sharp image
FINITE (M)   [-FIN-]
having limits or bounds, as numbers or magnitudes
INFINITE (M)   [IN-(not) + -FIN-]
always exceeding in value any assigned number

*(confine, define, definitive, final, finale, fine, finish, infinitive)*


## -FLECT-, -FLEX- (bend)


DEFLECTION  (P)   [DE-(away) + -FLECT-]
deviation from zero of the indicating pointer of a measuring instrument; bending of radiation from a straight path
FLEXURAL (STRENGTH) (P)   [-FLEX-]
resistance to bending
REFLECTANCE (P)   [RE-(back) + -FLECT-]
proportion of incident light reflected from a surface
REFLEX (M)   [RE-(back) + -FLEX-]
of an angle greater than a straight angle

*(circumflex, deflect, flex, flexible, inflect, reflective,
reflexive)*


## -FLU- (flow)


FLUORESCENCE  (P)  ["FLUORite" (mineral exhibiting
this property) < -FLU-]
property of substances emitting visible light while acted
upon by radiant energy; distinguished from PHOSPHORES-
CENCE
FLUORINE (C)   ["FLUORite" < -FLU-]
element (so named from its occurrence in the mineral
FLUORITE, which is used as a flux in steel and glassmaking)

FLUX (P)   [-FLU-]

rate of flow of fluids, particles, energy over a surface; in metallurgy, a substance that helps bonding of metals through cleaning action

REFLUX (C)   [RE-(back) + -FLU-]

to heat in an apparatus so that the escaping vapor is continually condensed and reheated

*(affluence, confluence, fluctuation, fluent, fluid, influx, superfluous)*

## -FOC- (central point)

CONFOCAL (M)   [CON-(together) + -FOC-]

having same focus or foci

FOCUS (M, P)   [Latin: focus (hearth: thus burning point of a lens)]

either of the two fixed points which determine a given curve (so named originally because the burning point or focus of a parabolic mirror is at the geometrical focus of its curvature); point at which light rays converge after refraction or reflection

*(focal infection, focalize, foyer)*

## -FORM- (form)

CONFORMAL (S)   [CON-(together) + -FORM-(form: thus having same form or shape)]

of a method of projection for maps which preserves the shape of the areas or objects depicted

DEFORMATION (P)   [DE-(from, away) + -FORM-]

change of form or shape because of stress or pressure

FORMULA (C, M)   [-FORM-]

combination of figures and symbols showing composition and structure of a compound; set of algebraic symbols expressing a rule or combination

TRANSFORM (M, P)   [TRANS-(across) + -FORM-]
    change an expression or operation into an equivalent form
    or one with similar properties; change one energy form
    into another, or change a current in potential or type

*(conformist, format, formative, informal, reformation, uniform)*

## -FORT- (strong)

AQUA FORTIS (C)   [-AQUA-(water, liquid) + -FORT-]
    early scientific and still popular name for commercial
    nitric acid (so named because it is a powerful solvent and
    corrosive)
FORCE (P)   [-FORT-]
    something that changes or is capable of changing the state
    of rest or motion in a body

*(comfort, discomfort, effort, enforce, forte, fortify, fortitude)*

## -FRACT- (break)

DIFFRACTION (P)   [DIF < DIS-(apart) + -FRACT-]
    modification (breaking up and scattering) of light rays by
    partial obstruction or in passing by edges of opaque
    bodies
FRACTION (C, M)   [-FRACT-]
    part separated from a substance by fractionation; numeri-
    cal quantity less than a unit or expressed as one or more
    aliquot parts of a unit
FRACTIONATE (C)   [-FRACT-]
    separate a mixture into component parts by successive
    operations of distillation, crystallization, etc.
REFRACTORY (C)   [RE-(back) + -FRACT-(break: thus
        resisting control, stubborn)]
    capable of enduring or resisting high temperature or
    chemical corrosion

*(fractious, fragile, fracture, infraction, infringe, irrefragable)*

**45**

## -FUG- (flee)

CENTRIFUGAL (FORCE) (P)   [-CENTRI-(center) +
    -FUG-]
inertial reaction tending to cause bodies to move away
from a center it revolves about; compare CENTRIPETAL
FORCE and CENTRIFUGE

FUGACITY (C)   [-FUG-]
corrected vapor pressure used as a measure of the tendency of a substance to escape or disappear from a heterogeneous system

*(febrifuge, fugitive, fugue, refuge, refugee, subterfuge)*

## -FUS- (pour)

DIFFUSION (P)   [DIF < DIS-(apart) + -FUS-]
gradual intermixture and spreading of two fluids by
thermal agitation; scattering of light rays by reflection
from an irregular surface or transmission through a translucent medium

EFFUSION (C)   [EF < EX-(out) + -FUS-]
flow of gases under pressure through porous bodies

FUSIBLE (C)   [-FUS-]
of a metal or alloy having a low melting point

FUSION (C, P)   [-FUS-(pour: thus often of two substances
    melting together or blending)]
change of a substance from solid to liquid state; thermonuclear reaction in which nuclei of light atoms are united
to form heavier nuclei

*(confuse, diffuse, effusive, infuse, profuse, refuse, suffuse,
                transfuse)*

## -GALACT- (milk)

GALACTOMETER (C)   [-GALACT- + -METER-
    (measure)]

46

hydrometer for measuring the specific gravity of milk; also called LACTOMETER

GALAXY (P) [GALAX < -GALACT-]
any large gravitational system of stars; when capitalized refers to our galaxy, the MILKY WAY

*(galactagogue, galax, galaxy, lactate, lacteal, lettuce)*

## -GALVAN- (electric)

GALVANIZE (C) ["Luigi Galvani" (early investigator of electricity developed by chemical action)]
coat metal with protective layer of zinc by dipping (so named because original coating process employed galvanic action)
GALVANOMETER (P) [-GALVAN- + -METER- (measure)]
instrument for detecting and measuring tiny electric currents

*(galvanic, galvanize, galvanothermy)*

## -GEL- (congeal)

GEL (C) [-GEL-]
solid phase of a colloidal solution, which is jellylike in consistency
REGELATION (P) [RE-(again) + -GEL-]
process of melting and freezing again when pressure is removed

*(congeal, gelatin, gelatinous, gelid, gelignite, jell, jelly)*

## -GEN- (produce)

DEGENERATION (P) [DE-(reversal) + -GEN-]
process of reducing signal strength and distortion in an

47

amplifier by feeding back output power into input circuit; opposed to REGENERATION

-GEN (C)    [-GEN-]

combining form in the names of chemical elements, such as HALOGENS ("salt producers," so named because they form a salt by direct union with a metal); HYDROGEN ("water producer," so named from the result of the combustion of hydrogen); OXYGEN ("acid producer," so named because originally considered essential for all acids)

GENERATRIX (M)    [-GEN-]

point, line or figure whose motion traces out another figure

HOMOGENEOUS (C, M)    [-HOMO-(same) + -GEN-(race, family)]

having uniform composition throughout; compare HETEROGENEOUS; of an equation having all terms of same degree

*(congenital, engender, generate, generic, genesis, indigenous, progeny)*

## -GEO- (earth)

APOGEE (P)    [APO-(from) + GEE < -GEO-]

greatest distance from earth of orbiting heavenly body or vehicle; opposed to PERIGEE

GEO- (P)    [-GEO-]

combining form meaning earth, as in GEOMAGNETIC, GEOPHYSICS, GEOTHERMAL

GEODESY (M)    [-GEO- + -DES-(divide)]

branch of applied mathematics which determines the figures and areas of large portions of the earth's surface

GEOMETRY (M)    [-GEO- + -METR-(measure)]

branch of mathematics dealing with relations, properties, and measurements of points, lines, angles, surfaces, and

48

solids (so named because originally the practical art of surveying and measuring land surfaces)

*(geode, geography, geology, geomancy, geophagy, geopolitics, georgic)*

## -GLOMER- (ball)

AGGLOMERATION (P)   [AG < AD-(to, into) + -GLOMER-]
clustering together of particles in a fluid by the action of sound waves

CONGLOMERATE (C)   [CON-(together) + -GLOMER-]
made up of separate substances gathered together into a single mass; in geology, type of rock consisting of fragments or pebbles cemented loosely in a mass

*(agglomeration, conglomeration, glomerate)*

## -GLYC- (sweet)

GLUCINUM (C)   [GLUC < -GLYC-]
former name for the element beryllium (so named because some of its salts are sweet to the taste)

GLYCEROL (C)   [-GLYC-]
sweet, syrupy, colorless alcohol; also GLYCERIN

*(none)*

## -GON- (angle)

DIAGONAL (M)   [DIA-(through) + -GON-]
straight line joining two nonadjacent angles of a figure

ISOGONIC (M, P)   [-ISO-(equal) + -GON-]
having equal angles; of a line connecting points on the earth's surface having equal magnetic declination (angle with true north)

POLYGON (M)   [-POLY-(many) + -GON-]
> closed plane figure bounded by straight lines, usually more than four; compare DECAGON, OCTAGON

TRIGONOMETRY (M)   [-TRI-(three) + -GON- + -METR-(measure)]
> branch of mathematics dealing with the relations of the sides and angles of triangles

*(gonion, Pentagon)*


## -GOVERN- (*see* -CYBERN-)


## -GRAD-, -GRESS- (step, go)

CENTIGRADE (C, P)   [-CENT-(100) + -GRAD-]
> of a temperature scale having 100 degrees between freezing and boiling points of water

GRADIENT (P)   [-GRAD-]
> rate of change of variable quantities such as temperature or pressure

GRADUATE (C)   [-GRAD-]
> container marked in units for measuring fluids or solids

PROGRESSION (M)   [PRO-(forward) + -GRESS-]
> series of numbers or quantities, each derived from preceding by a constant principle

*(aggression, congress, degrade, digress, gradual, ingredient, transgress)*


## -GRAM- (writing)

GRAM (C, P)   [Latin: gramma (a small weight, from the marking thereon) < Greek: GRAMma (writing)]
> basic unit of mass or weight in the cgs system; also in combinations as CENTIGRAM, KILOGRAM

-GRAM (P)   [-GRAM-]
  combining form meaning written or drawn, as in CHRONO-
  GRAM, OSCILLOGRAM

PARALLELOGRAM (M)   [-PARALLEL-(parallel) +
    -GRAM-(letter: thus stroke in writing)]
  four-sided plane figure with opposite sides parallel and
  equal

PROGRAMMING (M)   [PRO-(before, i.e., before the pub-
    lic) + -GRAM-]
  process of setting up a procedure for problem solving
  which can be read by a specific computer

  *(anagram, diagram, epigram, glamour, grammar, telegram)*


### -GRAPH- (write)


GRAPH (C, M, P)   ["GRAPHic formula" < -GRAPH-]
  diagram representing varying relationships between two
  or more factors by visual means; also, in mathematics,
  locus of all points whose coordinates satisfy the functions
  of an equation

-GRAPH (P)   [-GRAPH-]
  combining form meaning that which writes or records, or
  the writing or record itself, as SEISMOGRAPH, SPECTROGRAPH

GRAPHITE (C)   [-GRAPH-]
  variety of carbon used for pencil leads, lubricants, etc.

ORTHOGRAPHIC (S)   [-ORTHO-(right angle) +
    -GRAPH-]
  of a projection using lines at right angles to plane of
  projection

  *(autobiography, autograph, graphic, homograph,*
  *paragraph, stenography)*

## -GRAV- (heavy)

GRAVIMETRIC (C)   [-GRAV- + -METR-(measure)]
    of an analysis according to weight; contrast VOLUMETRIC
GRAVITATION (P)   [-GRAV-]
    force whereby every particle in the universe attracts every
    other particle; GRAVITY is terrestrial gravitation, the tend-
    ency of bodies to fall toward the center of the earth

    *(aggravate, aggrieve, grave, gravid, grief, grievous)*

## -GREG- (flock)

AGGREGATE (M)   [AG < AD-(to: thus bring to) +
    -GREG-]
    total of all points or numbers that satisfy a given condition
SEGREGATE (C)   [SE-(apart from) + -GREG-]
    separate from a mass and collect together in a particular
    place, as in solidifying or crystallizing

*(aggregation, congregate, egregious, gregarious, segregation)*

## -GRESS- (*see* -GRAD-)

## -GYR(O)- (rotating)

GYROMAGNETIC (P)   [-GYRO- + "magnetic"]
    referring to the magnetic properties of rotating electric
    charges, especially of electrons in atoms
GYROSTATICS (P)   [-GYRO- + -STAT-(stand)]
    branch of physics dealing with the laws of rotation of
    solid bodies

    *(girasol, girandole, gyrate, gyre, gyrocompass, gyro pilot,*
                             *gyroscope)*

## -HAB- (have, hold)

HABIT (C)   [Latin: habitus (what one customarily has or holds, thus condition, appearance) < HABere (have)]
   characteristic forms of crystallization developed by any one mineral

INHIBITOR (C)   [IN-(in) + HIB < -HAB-(hold: thus hold back, restrain)]
   substance that slows down or stops a chemical reaction; contrast CATALYST

*(dishabille, exhibit, habitual, inhibit, inhibition, prohibit)*


## -HEDR- (surface)

-HEDRAL (C, M)   [-HEDR-]
   combining form meaning having the specified number of faces or sides, as DIHEDRAL, POLYHEDRAL

-HEDRON (C, M)   [-HEDR-]
   combining form meaning a geometric figure or crystal with a specific number of faces or sides; as POLYHEDRON, TETRAHEDRON

*(cathedral, ex cathedra)*


## -HELI- (sun)

HELIUM (C)   [-HELI-]
   element (so named because first detected in the sun's spectrum)

PERIHELION (P)   [PERI-(around, near) + -HELI-]
   point nearest the sun in orbit of planet or comet; opposed to APHELION

*(heliacal, heliocentric, heliograph, heliotrope, heliotropism)*

# HEMI- (half)

HEMIHYDRATE (C)   [HEMI- + "hydrate" < -HYDR-
   (water)]
   hydrate in which there are half as many molecules of
   water as there are of the substance combined with the
   water
HEMISPHERE (M)   [HEMI- + "sphere" < -SPHER-
   (ball)]
   one of two half spheres obtained by a plane passing
   through a sphere's center

   *(hemialgia, hemicycle, hemidemisemiquaver, hemicycle,
   hemipterous)*

# -HES-, -HER- (stick)

ADHESION (P)   [AD-(to) + -HES-]
   force holding together molecules of dissimilar substances
   in contact
COHESION (P)   [CO-(together) + -HES-]
   force holding together molecules of same kind or same
   body

   *(adhere, adherent, coherent, hesitate, inhere, inherent)*

# -HETERO- (different)

HETERODYNE (P)   [-HETERO- + -DYN-(power)]
   combine radio waves of different frequencies to produce
   beat frequencies
HETEROGENEOUS (C)   [-HETERO- + -GEN-]
   having non-uniform composition throughout; compare
   HOMOGENEOUS

   *(heterodox, heterogeneity, heterography, heteronym,
   heterosexual)*

## -HEX(A)- (six)

HEX(A)-(C)   [-HEX(A)-]
   combining form indicating six atoms, molecules, groups, etc., as HEXAHYDRATE, HEXOSE
HEXAGON (M)   [-HEXA- + -GON-(angle)]
   polygon with six sides and angles

   *(hexaemeron, hexagram, hexameter, hexapla, Hexateuch)*


## -HOL(O)- (whole, complete)

HOLO- (C)   [-HOLO-]
   combining form indicating highest possible number of hydroxyl groups, as HOLOPHOSPHORIC ACID
HOLOMORPHIC (C)   [-HOLO- + -MORPH-(form)]
   of crystals whose opposite ends are wholly symmetrical in form

   *(holocaust, Holocene, holograph, holism, holophote, holophrastic)*


## -HOMO- (same)

HOMOGENEOUS (C, M)   [-HOMO- + -GEN-(kind)]
   having uniform composition throughout; compare HET-EROGENEOUS; of an equation having all terms of same degree
HOMOLOGOUS (C)   [-HOMO- + -LOG-(proportion)]
   of a series of compounds whose members differ from adjacent members by a constant increment

   *(anomalous, homogenized, homograph, homonym, homophone, homosexual)*

## -HYDR(O)- (water)

ANHYDROUS (C)  [AN-(without) + -HYDR-]
  of a compound without water or water of crystallization
  in its composition
DEHYDRATE (C)  [DE-(from) + -HYDR-]
  remove water from
HYDR(O)- (C)  [-HYDR(O)-]
  combining form meaning water, as in HYDRATE, HYDRAU-
  LICS, HYDROLYSIS; or compound of hydrogen, as in HYDRO-
  CARBON, HYDROCHLORIC, HYDROGENATE
HYDROGEN (C)  [-HYDRO- + -GEN-(producer)]
  chemical element (so named from the result of the combustion of hydrogen)

*(dropsy, hydra, hydrangea, hydrant, hydrocephalous,
hydrophobia, hydrofoil)*

## -HYGR(O)- (moisture)

HYGROMETER (P)  [-HYGRO- + -METER-(measure)]
  instrument for measuring the amount of moisture in the
  air
HYGROSCOPIC (C)  [-HYGRO- + -SCOP-(observe: thus
  indicating roughly the presence or absence of humid-
  ity)]
  absorbing or condensing moisture from the air; compare
  DELIQUESCENT

*(none)*

## HYPER- (above, beyond)

HYPERBOLA (M)  [HYPER- + -BOL-(throw: thus a throw-
  ing beyond, excess)]
  curve formed by the intersection of a plane with a right

circular cone (so named because the inclination of the plane to the base of the cone exceeds that of the side of the plane)

HYPERSONIC (P)   [HYPER- + "sonic" < -SON-(sound)]
   of speeds greater than speed of sound, Mach 5 or greater

   *(hyperbole, hypercorrection, hypercritical, hyperdulia, hypertension)*

## HYPO-  (under, below)

HYPO- (C)   [HYPO-]
   prefix signifying lowest degree of oxidation in a series of compounds, as HYPOCHLOROUS ACID
HYPOCYCLOID (M)   [HYPO- + -CYCL-(circle)]
   curve traced by a point on a circle rolling on the inside of a fixed circle; compare EPICYCLOID
HYPOTENUSE (M)   [HYPO- + -TEN-(stretch)]
   side of a right triangle under or opposite the right angle
HYPOTHESIS (S)   [HYPO- + -THESIS-(place: thus foundation)]
   assumed explanation of observed data or facts

   *(hyphen, hypocenter, hypochondria, hypodermic, hypothetical, hypothyroidism)*

## -IENT- ( see -IT-)

## IN- (in, into)
### ( also IL-, IM-, IR-, etc., depending on following letter)

IMPEDANCE (P)   [IM < IN-(in) + -PED-(foot: literally, entangling the feet, and thus hindering)]
   in a circuit, total opposition to flow of alternating current; in a sound-transmitting medium, ratio of force per unit area to volume displacement of a specified surface

57

INDUCTION (P)  [IN- + -DUCT-(to lead)]
    production of electrical or magnetic effects in a conductor
    or magnetizable body in proximity to the influence or
    variation of a field of force
INHIBITOR (C)  [IN- + HIB < -HAB-(hold: thus hold
        back, restrain)]
    substance that slows down or stops a reaction; compare
    CATALYST
IRRADIATION (P)  [IR < IN- + -RADI-(ray)]
    exposure to radiation, or amount of radiation incident on
    a surface at a given time

    (illumine, immigrate, indent, infer, inherent, inspire,
                involve, inundate)


## IN- (not)
### (also IL-, IM-, IR-, etc., depending on following letter)

IMMISCIBLE (C)  [IM < IN- + -MISC-(mix)]
    of liquids incapable of being mixed to produce a homo-
    geneous substance
INDETERMINATE (M)  [IN- + DE-(intensive, thus "com-
        pletely") + -TERM-(limit)]
    unlimited as to number of solutions or fixed values
INFINITE (M)  [IN- + -FIN-(limit)]
    always exceeding in value any assigned number
IRREGULAR (M)  [IR < IN- + "regular" < -REGUL-
        (rule)]
    of a polygon whose sides and angles are not equal

    (illegible, immutable, incognito, infamy, innocuous,
                irrational, insipid)


### -INDIC- (point out)

INDICATOR (C)  [-INDIC-]
    substance which by color change shows change in or con-
    dition of a system

INDEX (M, P)   [INDEX < -INDIC-]
> subscript or superscript to indicate position in an arrangement or operation to be performed; INDEX OF REFRACTION is the ratio of the velocity of light, etc. in a vacuum to its velocity in a specific medium

## INFRA- (below)

INFRARED (P)   [INFRA- + "red"]
> of wavelengths just below or beyond the visible red spectrum

INFRASONIC (P)   [INFRA- + "sonic" < -SON-(sound)]
> of those frequencies below audible sound

> *(infra dig, infrahuman, infrapose, infrastructure)*

## -INSUL- (set apart)

INSULATOR (P)   [Latin: insula (island: thus set apart)]
> body of nonconducting material for separating and supporting charged conductors

ISOLATE (C)   [ISOL < -INSUL-]
> separate from other substances and obtain free or uncombined

> *(insular, insulin, isle, isolable, isolationist)*

## INTEG- (whole)

INTEGER (M)   [-INTEG-]
> whole number

INTEGRAL (M)   [-INTEG-]
> limit or whole sum of the series of values assumed by a differential $f(x)dx$ when x varies by indefinitely small increments

DISINTEGRATION (P)   [DIS-(apart, reversal) +
    -INTEG-]
    change in the structure of an atomic nucleus, as through
    emission of nucleons by a radioactive element
INTEGRATOR (P)   [-INTEG-]
    device which continually totals a quantity measured dur-
    ing a period of time

*(entire, integer vitae, integral, integration,
integrity, redintegrate)*


**INTER-** (between)


INTERCEPT (M)   [INTER- + -CEPT-(take)]
    cut off or include between two points, lines, or planes
INTERFACE (P)   [INTER- + -FACE-(surface)]
    surface forming boundary between adjacent parts of mat-
    ter or space
INTERPOLATE (M)   [INTER- + -POLA-(calculate)]
    determine intermediate values between known values;
    compare EXTRAPOLATE
INTERSECTION (M)   [INTER- + -SECT-(cut)]
    point or line common to two or more lines, planes, or
    surfaces

*(intercede, interfere, interjection, intermit,
interstices, intervene)*


**-ION-** (go)


ION (C, P)   [-ION-]
    electrically charged atomic or molecular particle resulting
    from the loss or gain of one or more electrons (so named
    by Faraday because they go to the poles or electrodes in
    electrolysis); note also ANION, CATION, THERMION

IONIUM (P)   [-ION- + "uranIUM"]
  radioactive isotope of thorium, produced by disintegration
  of uranium (so named from its ionizing activity)

*(none)*

## -IRID- (rainbow)

IRIDESCENCE (C, P)   [-IRID-]
  interplay of rainbowlike colors on the surface of a substance
IRIDIUM (C)   [-IRID-]
  element (so named from the iridescence of some of its
  salts)

*(Iris, iris)*

## -ISO- (equal)

ISOMER (C, P)   [-ISO- + -MER-(part)]
  one of two or more compounds with same composition but
  different atomic arrangements; one of two or more nuclear
  species with the same number of neutrons and protons but
  differing in energy characteristics and radioactive properties
ISOMETRIC (P, S)   [-ISO- + -METR-(measure)]
  of a line on a diagram indicating pressure or temperature
  changes in a gas at constant volume; of a projection in
  which dimensions are shown in actual measurement and
  not in perspective
ISOSCELES (M)   [-ISO- + -SCEL-(leg)]
  of a triangle having two equal sides
ISOTOPE (P)   [-ISO- + -TOP-(place)]
  one of two or more forms of an element having same
  atomic number and position in periodic table and similar
  chemical properties but different atomic weights

*(isocracy, isometrics, isonomy, isopyre)*

## -IT-, -IENT- ( go )

AMBIENT (C, P)    [AMB(I)-(around) + -IENT-]
of the surrounding medium, as AMBIENT TEMPERATURE

CIRCUIT (P)    [CIRCU(M)-(around) + -IT-]
complete electric or magnetic path

TRANSIENT (P)    [TRANS-(across) + -IENT-]
brief electrical oscillation in a circuit caused by voltage or load change

TRANSITION (C, P)    [TRANS-(across) + -IT-]
of the three triads of elements forming Group 8 of the periodic table (so named because each group of three is the connecting link between two periods); sudden change in the energy state of an atom or nucleus

*(ambition, concomitant, exit, initial, sedition, transitive, transitory)*

## -JECT-( throw )

OBJECTIVE (P)    [OB-(against) + -JECT-]
lens or lens system that is nearest the object observed

PROJECT (M)    [PRO-(forward) + -JECT-]
represent a solid on a plane through lines of correspondence

*(abject, conjecture, deject, eject, interject, reject, subject, trajectory)*

## -JUG-, -JUNCT- ( join )

CONJUGATE (C, M)    [CON-(together) + -JUG-]
having two or more radicals acting as one; reciprocally related and interchangeable as to certain properties, such as two points, lines, quantities

THERMOJUNCTION (P)  [-THERM-(heat) + -JUNCT-]
 point of contact of a pair of conductors forming a thermocouple

> *(adjunct, conjugal, conjunction, disjunctive,*
> *injunction, juncture)*

### -KINE- (movement)

KINEMATICS (P)  [-KINE-]
 branch of mechanics dealing with motion in the abstract
 without reference to forces or masses
KINETICS (P)  [-KINE-]
 branch of physics dealing with the motion of bodies in
 relation to the forces acting on them

> *(cinema, kinescope, kinesthesia, telekinesis)*

### -LABOR- (work)

COLLABORATE  (S)  [COL<COM-(together) +
   -LABOR-]
 work with another in scientific endeavor
LABORATORY (S)  [-LABOR-]
 room or building where the scientist works

> *(belabor, elaborate, labor, laborious)*

### -LANTH-, -LAT- (concealed)

LANTHANUM (C)  [-LANTH-]
 element (so named because it had lain concealed in oxide
 of cerium until its discovery)
LATENT (HEAT) (P)  [-LAT-]
 amount of additional heat needed to convert a solid into
 liquid or liquid into vapor after reaching melting or boiling point (so named because it did not raise the tempera-

ture or could not be felt as warmth, and thus was considered as being absorbed and remaining latent in the resulting liquid or vapor)

*(latent, latency, latescent)*

### -LAT- (carry) (*see*-FER-)

### -LAT- (wide)

DILATOMETER (P)  [DI(S)-(apart) + -LAT- + -METER-(measure)]
  instrument for measuring the expansion of liquids or solids
LATITUDE (P)  [-LAT-]
  angular distance of a celestial body from the ecliptic; in geography, angular distance north or south of the equator

*(dilatation, dilate, dilative, latifundium, latitudinal, latitudinarian)*

### -LATER- (side)

EQUILATERAL (M)  [-EQUI-(equal) + -LATER-]
  of a figure having all sides equal
LATUS RECTUM (M)  [Latin: latus(side) + rectum (straight)]
  straight line through the focus of a conic at right angles to the transverse diameter

*(collateral, lateral, laterigrade)*

### -LEV- (raise)

LEVER (P)  [-LEV-]
  device consisting of a bar pivoting on a fulcrum to impart

force applied at one point to a resisting force at another point

ELEVATION (P)  [E-(out, up) + -LEV-]
    angular distance of any heavenly body above the horizon

*(elevate, elevator, leaven, Levant, levee,
levitate, levy, relevant)*

### -LIBR- (balance)

EQUILIBRANT (P)  [-EQUI-(equal) + -LIBR-]
    force or forces necessary to balance or keep in equilibrium another force or forces

EQUILIBRIUM (C, P)  [-EQUI-(equal) + -LIBR-]
    state of even balancing between rates of forward and reverse reactions, or between forces whose resultant is zero

*(deliberate, equilibrate, equilibrist, level, lb.,
Libra, librate)*

### -LIG- (bind)

ALLOY (C)  [AL < AD-(to) + LOY < LEI < -LIG-]
    mixture obtained by fusion of two or more metals

COLLIGATIVE (C, P)  [COL < COM-(together) + -LIG-]
    of physical or chemical properties related by a mathematical function

*(ally, league, liable, lien, ligament, ligature,
oblige, religion, rely)*

### -LIM- (threshold, lintel)

ELIMINATE (M)  [E-(out) + -LIM-(threshold: thus to put out of doors, expel)]
    remove an unknown by combining equations

SUBLIMATE (C)  [SUB-(up to) + -LIM-(lintel: thus to raise up)]
    change from solid to gaseous state and back to solid without apparent liquefaction

        *(eliminate, lintel, preliminary, sublimation, sublime, subliminal)*


## -LIN- (line)

ALIGN (P)  [A(D)-(into) + -LI(G)N-]
    adjust elements of electronic equipment for best performance

COLLIMATE (P)  [COL < COM-(together) + LIM < -LIN-]
    make rays of light parallel; adjust line of sight of an optical instrument

COLLINEAR (M)  [COL < COM-(together) + -LIN-]
    lying in the same straight line; note also CURVILINEAR, RECTILINEAR

LINEAR (M, P)  [-LIN-]
    of an equation whose variables are in the first power only, and thus plots as a straight line; of an output directly proportional to input

    *(delineate, line, lineage, lineal, lineament, lineate)*


## -LIPS- (leave)

ECLIPSE (P)  [EC < EK-(out) + -LIPS-(leave: thus pass over, fail)]
    partial or total elimination of the light of one heavenly body by another; note also ECLIPTIC (so named because eclipses occur on this orbit)

ELLIPSE (M)  [EL < EN-(in, behind) + -LIPS-(leave: thus fall short, be lacking)] curve formed when a cone is cut obliquely by a plane making a smaller angle with the base than the side of the cone makes with the base (so named because the inclination of the plane to the base falls short of or is less than that of the side of the cone)

*(ellipsis)*

## -LITH- (stone)

LITHARGE  [-LITH- + -ARG-(silver)]
   lead monoxide (so named because obtained as a by-product of smelting silver-bearing ores)

LITHIUM (C)  ["LITHia"(lithium oxide) < -LITH-(stone: name was proposed to designate this alkali's mineral origin)]
   element

*(Eolithic, lithograph, lithosphere, lithotomy, megalith, monolithic, oolite)*

## -LIQU- (liquid)

DELIQUESCE (C)  [DE-(an intensive) + -LIQU-]
   become liquid through absorbing moisture from the air

LIQUID (C, P)  [-LIQU-]
   not solid or gaseous; compare FLUID; note also LOX (liquid oxygen)

*(liquefy, liqueur, liquidate, liquor)*

## -LOC- (place)

LOCAL (ACTION) (C)  [-LOC-]
   voltaic action produced in one of the plates of an electrolytic cell because of impurities in the plate

LOCUS (M)   [-LOC-]
> surface or curve considered as traced by a point or line according to given conditions (so named because it traces the places where the point or line can be)

> *(allocate, collocate, dislocate, lieu, lieutenant, locate, locomotion)*

## -LOG- (proportion, word)

ANALOG (M)   [ANA-(according to) + -LOG-(proportion)]
> class of computers using numbers represented by measurable physical quantities; compare DIGITAL COMPUTER using numbers expressed directly as digits

HOMOLOGOUS (C)   [-HOMO-(same) + -LOG-(proportion)]
> of a series of compounds whose members differ from adjacent members by a constant increment

LOGARITHM (M)   [-LOG-(proportion) + -ARITHM-(number)]
> power to which a fixed number, or base, must be raised to produce a given number

-LOGY (S)   [-LOG-(word, discourse: thus extended to body of knowledge from which one speaks)]
> combining form meaning science or study of, as COSMOLOGY, METEOROLOGY, TECHNOLOGY

> *(analogy, dialogue, eulogy, logic, logorrhea, logos, monologue, prologue)*

## -LUC- (shine)

LUX (P)   [LUX < -LUC-]
> unit of illumination in the metric system

TRANSLUCENT (P)   [TRANS-(across) + -LUC-]
    allowing passage of light but an unclear view of an object;
    compare TRANSPARENT

    *(elucidate, lucid, lucifer, lucite, lucubration, pellucid)*


## -LUMIN- (light)

ILLUMINANCE (P)   [IL < IN-(in) + -LUMIN-]
    intensity of light per unit area of an evenly illuminated
    surface
LUMEN (P)   [LUMEN < -LUMIN-]
    unit of measure for time rate of flow of visible light
LUMINESCENCE (P)   [-LUMIN-]
    giving off of visible light that is not due to heat of incan-
    descence; as FLUORESCENCE, RADIOLUMINESCENCE, THERMO-
    LUMINESCENCE
LUMINOSITY (P)   [-LUMIN-]
    measured brightness of a color or surface

    *(illuminate, illuminati, limn, luminary,
    luminiferous, luminous)*


## -LUN- (moon)

LUNE (M)   [-LUN-]
    crescent-shaped figure formed by two arcs of circles
CISLUNAR (P)   [CIS-(on this side) + -LUN-]
    between earth and moon

    *(lunacy, lunatic, lunar, lunarian, lunate,
    lunette, lunisolar)*


## -LUT- (wash)

DILUTE (C)   [DI(S)-(away) + -LUT-(wash: thus make
    less concentrated)]
    weaken or thin by addition of liquid such as water

ELUTRIATE (C)   [E-(out) + -LUT-]
   separate lighter particles from heavier by washing and
   decanting

   *(ablution, alluvial, antediluvian, deluge)*

## -LYS-, -LYT- (loosen)

ANALYSIS (C, M)   [ANA-(up) + -LYS-]
   determining nature or proportion of constituent parts of
   a substance by separating the ingredients; working out of
   problems by means of equations or examining the relations of variables
CATALYSIS (C)   [CATA-(an intensive) + -LYS-]
   speeding up of reaction by adding substance (CATALYST)
   which remains unchanged
ELECTROLYTE (C)   [-ELECTR-(electric) + -LYT-
      (loosen, dissolve)]
   substance which in solution dissociates into free ions and
   conducts electricity
LYSIMETER (C)   [-LYS-(loosen, dissolve) + -METER-
      (measure)]
   instrument for determining solubility of substances

   *(analytic, LSD, lysol, lysis, palsy, paralysis, psychoanalysis)*

## -MACR(O)- (large)

MACRO- (C, P)   [-MACRO-]
   combining form meaning on a large scale; often contrasted
   with micro-compounds, as MACROANALYSIS, MACROMOLE-
   CULE, MACROSCOPIC
MACROSONICS (P)   [-MACRO- + "sonics" < -SON-
      (sound)]
   technology of high power sound

   *(macroclimate, macrocosm, macrograph, macron)*

## -MAGNET-, -MAGNES- ( magnetic )

MAGNESIUM (C)   [Latin: magnesia alba (magnesia white
    or magnesium carbonate) < magnes carneus (flesh
    magnet) < magnes (lodestone) < Magnesia lithos
    (mineral of Magnesia, a district in Thessaly)]
    element (so named because it is the base of magnesia
    (originally hydrated magnesium carbonate), called mag-
    nesia white to distinguish it from magnesia black or man-
    ganese dioxide; the name magnesia white seems to have
    been suggested by magnes carneus, a white earth that
    adhered strongly to the lips and thus was supposed to
    have the same attraction for flesh that the lodestone had
    for iron)

MAGNET (P)   [Latin: magnes, magnetis (lodestone or mag-
    netic iron oxide) < Greek: Magnesia lithos (mineral of
    Magnesia)]
    body which attracts iron and creates a magnetic field
    external to itself

MAGNETO- (C, P)   [-MAGNET-]
    combining form meaning magnetic or magnetism, as MAG-
    NETOCHEMISTRY, MAGNETOELECTRICITY, MAGNETOSTRICTION
    (-STRICT-"deformation")

MANGANESE (C)   [corruption of Latin: magnesia (man-
    ganese dioxide) < Greek: Magnesia lithos (mineral of
    Magnesia)]
    element (so named because it is the base of magnesia
    black or manganese dioxide, called magnesia because of
    early confusion with magnes or lodestone, the original
    Magnesian mineral)

*(magnetic, magnetism, magneto, milk of magnesia)*

## -MAGNI- ( great, large )

MAGNIFICATION (P)   [-MAGNI- + -FIC-(make)]
    ratio of the dimensions of the image to those of the object

71

## MAGNITUDE (M, P)   [-MAGNI-]

number given to a quantity so that relative values of quantities of the same class may be compared; size of angle; relative brightness of stars expressed on a logarithmic scale

*(magistrate, magnanimous, magnate, magnificent, magniloquence, magnum)*

## -MAN- (remain)

## REMAINDER (M)   [RE-(back) + -MA(I)N-]

quantity left over after subtraction or addition

## REMANENCE (P)   [RE-(back) + -MAN-]

magnetic induction remaining in a magnetized material after the external magnetizing force is removed

*(immanent, impermanence, manor, manse, mansion, permanent, remnant)*

## -MEA- (pass)

## PERMEABILITY (P)   [PER-(through) + -MEA-]

measure of the rate of diffusion of a gas through a standard fabric under standard conditions; property of being readily traversed by magnetic flux

## PERMEANCE (P)   [PER-(through) + -MEA-]

ability to be traversed by magnetic flux; reciprocal of RELUCTANCE

*(impermeable, permalloy, permeate)*

## -MED- (middle)

## MEAN (M)   [French: meien < Latin: medianus < MEDius (middle)]

quantity whose value is intermediate between the values

of two or more quantities, as ARITHMETIC and GEOMETRIC MEAN; the second or third term in a four-term proportion

MEDIAN (M) [-MED-]
straight line extending from any vertex of a triangle to the middle of the opposite side; value of the middle point in a series of values

MEDIUM (P) [-MED-]
intervening substance through which a force or radiation is transmitted

MERIDIAN (P) [MER < -MED- + Latin: dies (day)]
great circle of the celestial sphere passing through the poles and the zenith of an observer at any place on the earth's surface (so named because the sun crosses it at noon)

*(immediate, intermediary, media, mediate, medieval, mediocre, Mediterranean)*

## -MEG(A)- (large)

MEGA- (S) [-MEGA-]
combining form in metric system and technical use meaning one million, as MEGACYCLE, MEGATON, MEGOHM

MEGASCOPIC (C) [-MEGA- + -SCOP-(observe)]
visible to the naked eye, synonym of MACROSCOPIC

*(megabuck, megalith, megalomania, megalopolis, megaphone)*

## -MENS- (measure)

COMMENSURABLE (M) [COM-(together) + -MENS-]
having a common measure, divisible by the same quantity without remainder

DIMENSION (M) [DI(S)-(off, from) + -MENS-]
one of the three coordinates required to determine position; sum of the exponents in a term

*(commensurate, immense, measure, measured)*

# -MER- (part)

ISOMER (C, P)   [-ISO-(equal) + -MER-]
  one of two or more compounds with same composition but different atomic arrangements; one of two or more nuclear species with the same number of neutrons and protons but differing in energy characteristics and radioactive properties
POLYMERIZATION (C)   [-POLY-(many) + -MER-]
  reaction which combines smaller molecules to form larger molecules with repeating structural units and same percentage composition

*(merism, trimerous)*

# -MERS- (dip)

EMERSION (P)   [E-(out) + -MERS-]
  reappearance of a celestial body after going behind or into the shadow of another
IMMERSION (P)   [IM < IN-(in) + -MERS-]
  disappearance of a celestial body by going behind or into the shadow of another

*(emergency, emergent, immerse, merganser, merge,
merger, submerge, submersion)*

# -MES- (middle)

MESOMORPHIC (P)   [-MES- + -MORPH-(form)]
  of a state of matter intermediate between the true liquid and crystal
MESON (P)   [-MES-]
  unstable particle whose mass is between that of the proton and electron

*(mesomorph, Mesopotamia, Mesozoic)*

74

## META- ( across, beyond )

METASTABLE (C, P)  [META- + "stable" < -STA-(stand)]
energized beyond the most stable state but in an apparent
state of equilibrium; of an atom in certain excited states
in which radiation does not occur and thus momentarily
stable

METATHESIS (C)  [META- + -THESIS-(place: thus
interchange)]
reaction involving interchange of elements or radicals
between two compounds: double decomposition

*(metabolism, metamorphosis, metaphor, metaphysics,
metempsychosis)*

## -METEOR- ( atmosphere )

METEOROID (P)  [-METEOR- < Greek: meteoron (thing
soaring in air)]
small solid body moving through space that becomes a
meteor upon entering earth's atmosphere; compare
METEORITE

METEOROLOGY  [-METEOR- + -LOGY-(study of)]
branch of physics dealing with weather and related
atmospheric phenomena

*(meteoric)*

## -METER-, -METR- ( measure )

DIAMETER (M, P)  [DIA-(through) + -METER-]
straight line which passes through the center of a circle or
sphere from one side to the other; unit of measurement of
the magnifying power of a lens (equal to the number of
times the linear size of the object is increased)

75

ISOMETRIC (P, S)   [-ISO-(equal) + -METR-]
> of a line on a diagram indicating pressure or temperature changes in a gas at constant volume; of a projection in which dimensions are shown in actual measurement and not in perspective

METER (C, P)   [-METER-]
> basic unit of length in the metric system, originally defined as one ten-millionth of the distance from the pole to the equator; note also metric units MILLIMETER, KILOMETER, etc.

-METER (C, P)   [-METER-]
> combining form meaning device or instrument for measuring, as DOSIMETER, HYDROMETER, THERMOMETER, VOLTMETER

-METRY (C, M, P)   [-METR-]
> combining form meaning process or science of measuring, as GEOMETRY, TELEMETRY, TRIGONOMETRY

PARAMETER (M)   [PARA-(beside) + -METER-]
> quantity or constant whose values characterize some varying member of a system of functions or expressions

PERIMETER (M)   [PERI-(around) + -METER-]
> length of a closed curve or sum of the sides of a plane figure

SYMMETRICAL (C, M)   [SYM-(together) + -METR-]
> showing a definite repeated arrangement of atoms in the structural formula; of an equation, expression, or relation whose terms are interchangeable without affecting its validity

*(asymmetry, diametrically, metrical, metronome, symmetry)*


### -MICRO- (very small)


MICRO- (C, P, S)   [-MICRO-]
> combining form meaning minute quantities or particles, as MICROCHEMISTRY, MICROPHYSICS (often contrasted with

macro-compounds); combining form in metric system and technical use meaning one millionth, as MICROFARAD, MICROGRAM, MICRON

MICROMETER (S) [-MICRO- + -METER-(measure)]
instrument for measuring small angles or dimensions

MICROSCOPE (C, P) [-MICRO- + -SCOP-(observe)]
instrument for enlarging small objects for examination

MICROWAVE (P) [-MICRO- + "wave"]
extremely short electromagnetic wave, especially between one and one hundred centimeters in wavelength; note also the acronym MASER, from the initial letters of "Microwave Amplification by Stimulated Emission of Radiation"

*(microbe, microcosm, microfilm, microgroove,
Micronesian, microorganism)*

### -MILL- (one thousand)

MIL (S) [-MIL(L)-]
unit of linear measure, equal to one thousandth of an inch

MILLI- (S) [-MILL-]
combining form in metric system and technical use meaning one thousandth, as MILLIAMPERE, MILLIMETER

*(mile, mill, millenary, millennium, million, millipede)*

### -MIN- (less)

COMMINUTE (C) [COM-(an intensive) + -MIN-]
reduce to fine particles, pulverize

MINUEND (M) [-MIN-]
number from which another is to be subtracted

MINUS (M) [-MIN-]
symbol denoting subtraction or negative quantity

MINUTE (M) [Latin: pars minuta prima (first small part:

77

i.e., the result of the first operation of sexagesimal division, earlier called prime minute, now simply "minute")
< MINus (less)]
unit of angular measure, 1/60 of a degree of arc

*(diminution, minimum, miniskirt, minister, minority,*
*minuet, minutiae)*

## -MISC- (mix)

IMMISCIBLE (C)   [IM < IN-(not) + -MISC-]
of liquids incapable of being mixed to produce a homogeneous substance
MIXTURE (C)   [MIXT < -MISC-]
substance consisting of two or more components which retain their own chemical properties; distinguished from COMPOUND

*(meddle, medley, miscegenation, miscellaneous,*
*miscellany, promiscuous)*

## -MISS-, -MIT- (send)

ADMITTANCE (P)   [AD-(to) + -MIT-]
measure of ease with which a circuit can carry an alternating current; reciprocal of IMPEDANCE
EMISSION (P)   [E-(out) + -MISS-]
giving off of energy or charged particles; note also the acronym LASER, from the initial letters of "Light Amplification by Stimulated Emission of Radiation"
PERMITTIVITY (P)   [PER-(through) + -MIT-]
dielectric constant, measure of the ability of a dielectric to hold electric potential
TRANSMIT (P)   [TRANS-(across) + -MIT-]
cause light, sound, heat, etc. to pass through a medium

*(demise, dismiss, emissary, intermittent, missile,*
*premise, remit, submit)*

# -MOD- (measure)

MODE (M)   [-MOD-]
    value having the greatest frequency in a frequency distribution

MODERATOR (P)   [-MOD-(measure: thus to measure by a standard, regulate)]
    substance used to slow down neutrons in a nuclear reactor

MODULATION (P)   [-MOD-(measure: thus to measure by a standard, regulate)]
    varying a characteristic of a carrier wave in accordance with another wave; note AM, FM

MODULUS (M, P)   [-MOD-]
    absolute value of a complex number (the numerical value of the length of the vector representing the complex number), or the factor by which natural logarithms are multiplied to convert to common logarithms; number or quantity that is the measure of a force or function

*(accommodate, commodious, discommode, model, modest, modicum, modify)*

# -MOL- (mass)

MOLAR (C, P)   [-MOL-]
    of a solution having one mole (gram-molecular weight) of solute per liter; of a body of matter considered as a whole

MOLECULE (C)   [-MOL- + -CUL (a diminutive ending: thus little mass)]
    smallest part of an element or compound that can exist separately and still retain its chemical properties

*(demolish, demolition, molest)*

## -MON(O)- (one)

MON(O)- (C)   [-MON(O)-]
   combining form meaning single, one: containing one atom
   of a specified kind per molecule (MONOCHLORIDE, MONOX-
   IDE); of acids having a single replaceable hydrogen atom
   (MONOBASIC); possessing a thickness of one molecule
   (MONOMOLECULAR)

MONOMIAL (M)   [-MO(NO)- + "biNOMIAL"]
   expression consisting of just one algebraic term

   *(monaural, monarchy, monolithic, monologue, monopoly,
   monotheism, monotony)*

## -MORPH- (form)

AMORPHOUS (C)   [A-(without) + -MORPH-]
   without definite shape or crystalline structure

POLYMORPHISM (C)   [-POLY-(many) + -MORPH-]
   condition of a substance having more than one form of
   crystallization with identical composition; distinguished
   from ISOMORPHISM

   *(anthropomorphic, metamorphosis, morpheme,
   Morpheus, morphine)*

## -MOV-, -MOT- (move)

ELECTROMOTIVE (P)   [-ELECTR-(electric) + -MOT-]
   of a force tending to bring about a flow of electricity
   between two points of different potential

MOMENT (P)   [Latin: momentum (movement) <
   movimentum < -MOV-]
   product of force or mass and its perpendicular distance
   from a point to the line of action of the force; tendency
   to cause rotation

MOMENTUM (P)   [Latin: momentum (movement) <
    movimentum < -MOV-]
    quantity of motion in a body, the product of its mass and
    velocity
PROMOTER (C)   [PRO-(forward) + -MOT-]
    substance which in small amounts will speed up the action
    of a catalyst

*(commotion, emotion, mob, mobile, motion, motive,
    motor, remote, remove)*


## -MULTI- (many)

MULTIPLE (M, P)   [-MULTI- + PLE(X) < -PLIC- (fold:
    thus more than one part)]
    any of the products of some specified number and another
    number; group of terminals arranged for electrical con-
    nection at many points
MULTIPLICATON (M)   [-MULTI- + -PLIC-(fold)]
    process of adding a number to itself a given number of
    times

*(multicolored, Multilith, multimillionaire, multiplicity,
    multitudinous)*


## -MUT- (change)

COMMUTATIVE (M)   [COM-(an intensive) + -MUT-]
    of an operation for which the order of the terms is
    irrelevant
MUTUAL (INDUCTION) (P)   [-MUT-(change, exchange)]
    induction produced in each other by two associated elec-
    tric circuits
PERMUTATION (M)   [PER-(an intensive) + -MUT-]
    process of rearranging elements in a series to achieve all
    possible changes of sequence, or any of the arrangements
    so produced

TRANSMUTATION (P)   [TRANS-(across) + -MUT-]
  conversion of one element into another by altering its
  nuclear structure

  *(commute, immutable, molt, mutable, mutant, transmute)*

## -NAT- (nature)

DENATURE (C, P)   [DE-(reversal) + -NAT-]
  change the nature of by making unsuitable for eating or
  drinking, as alcohol; make fissionable material unsuitable
  for military use by addition of non-fissionable material
NASCENT (C)   [Latin: NASCere, natus (be born: thus at
    moment of formation > NATura (nature)]
  of the state of an element just released from a compound
  and whose atoms are unusually active
NATIVE (C)   [-NAT-]
  of an element occurring in nature in a pure condition
NATURAL (SCIENCES) (S)   [-NAT-]
  sciences dealing with nature and the physical world, as
  BIOLOGY, CHEMISTRY, PHYSICS

  *(cognate, impregnate, innate, natal, nation, nativity,
      pregnant, supernatural)*

## -NE(O)- (new)

NEODYMIUM (C)   [-NEO- + "diDYMIUM"]
  element (so named when discovered as a component of
  didymium, formerly considered one of the elements)
NEON (C)   [-NE-]
  element, one of the "new" gases discovered in the nine-
  teenth century

  *(neoclassic, neologism, neomycin, neoorthodoxy,
      neophyte, neoprene)*

## -NEUTR- (neither)

NEUTRAL (C, P)  [-NEUTR-]
   neither acid nor alkaline; neither positive nor negative
NEUTRON (P)  [-NEUTR-]
   electrically uncharged particle of the atomic nucleus

*(neuter, neutrality, neutralize)*


## -NITR- (nitrogen)

NA (C)  [late Latin: NAtrium (sodium) < natron, nitron
   (niter, or native sodium carbonate)]
   symbol for the element sodium (the symbol for sodium
   and the element nitrogen both derive from niter, because
   in early use niter referred to native sodium carbonate as
   well as saltpeter, potassium nitrate)
NITROGEN (C)  [Greek: nitron (niter or potassium nitrate)
   + -GEN-(produce: thus literally niter producer)]
   element; note nitrogen compounds NITRATE, NITRIC, and
   combining form NITRO-, indicating the nitro group

*(none)*


## -NOD- (knot, loop)

ANTINODE (P)  [ANTI-(opposite to) + -NOD-]
   point midway between the nodes of a vibrating body
NODE (M, P)  [-NOD-]
   point where a curve crosses itself (CRUNODE) or turns
   sharply back on itself (ACNODE); point or line of complete
   or comparative rest in a vibrating body

*(denouement, nodule, nodose, noose)*

## -NOM- (law)

ASTRONOMY (P) [-ASTRO-(star, heavens) + -NOM-(law: thus science and study of)]
> science of the heavenly bodies: size, history, motion, constitution, etc.

NOMOGRAPH (M) [-NOM-(law: thus orderly arrangement) + -GRAPH-(write)]
> graph with graduated scales for interrelated variables, used to find unknown in relation to the known

*(antinomy, autonomous, economy, metronome, nomism, nomology)*

## -NOMI(N)- (name)

BINOMIAL (M) [BI-(two) + -NOMI-]
> algebraic expression consisting of two terms connected by a plus or minus sign

DENOMINATOR (M) [DE-(an intensive) + -NOMIN-]
> term in a fraction naming the number of equal parts into which the unit is divided; compare NUMERATOR

*(denominate, ignominy, nomenclature, nominal, nominate, noun, renown)*

## NON- (not)

NONCONDUCTOR (P) [NON- + "conductor" < CON-(together) + -DUCT-(to lead)]
> substance offering resistance to the passage of certain forms of energy, as sound, light, and particularly electricity

NONFERROUS (C) [NON- + -FERR-(iron)]

84

not containing any iron (usually refers to all metals except iron, steel, and alloys containing iron)

(*nonchalant, nondescript, nonentity, nonpareil, nonplus, non sequitur*)

## -NORM- (rule, standard)

NORM (M) [Latin: norma (carpenter's square: thus pattern, standard)]
mode, the value having the greatest frequency in a frequency distribution (thus the pattern or standard)

NORMAL (C, M) [-NORM-]
of a standard solution having one gram equivalent weight of the dissolved substance, used as a unit; perpendicular, especially a perpendicular to tangent or tangent plane at any point of a curve or curved surface (so named from original sense of carpenter's square)

(*abnormal, enormous, normality, normalize, normative*)

## -NUCL- (kernel)

NUCLEON (C, P) [-NUCL-]
any particle making up an atomic nucleus, as the NEUTRINO, NEUTRON, PROTON

NUCLEUS (C, P) [-NUCL-]
in organic chemistry, a stable arrangement of atoms that remains intact through a succession of chemical changes; central part of an atom containing nucleons that supply its effective mass and positive charge; bright central core of comet or nebula

(*enucleate, nucleate,*)

## -NUMER- (number)

NUMERATOR (M) [-NUMER-]
    term in a fraction expressing how many parts of a unit are to be taken; compare DENOMINATOR

NUMERICAL (M) [-NUMER-]
    indicating a value or magnitude without respect to sign

*(enumerate, innumerable, numeral, numeration,*
*numerous, supernumerary)*


## OB- (against)

OBTUSE (M) [OB- + -TUS-(beat: thus blunted)]
    of an angle greater than 90° but less than 180°

OCCLUSION (C) [OC < OB- + -CLUS-(shut)]
    taking up and retention, usually internally, of a gas by a solid

*(obdurate, objection, obloquy, obnoxious, obstacle,*
*obtuse, occur)*


## -OCT- (eight)

OCTANE (C) [-OCT-]
    one of a group of hydrocarbons with eight carbon atoms

OCTANT (M) [-OCT-]
    any of the eight areas into which a space is divided by three orthogonal planes intersecting at a single point

*(octave, octavo, octet, October, octogenarian,*
*octagonal, octopus)*

## -OD- (way)

CATHODE (C, P)   [CAT(A)-(down) + -(H)OD-]
  negative electrode, through which current leaves a non-
  metallic conductor; opposed to ANODE ("up way"); also
  electronic tube electrode emitting electrons
DIODE (P)   [DI(A)-(separation) + -OD-]
  two-element electron tube which permits electrons to
  move in one direction only
PERIOD (P)   [PERI-(around) + -OD-(way: thus a going
    around, cycle)]
  interval of time between two successive phases of an
  oscillation or any regularly recurring cyclical motion
PERIODIC (TABLE) (C)   [PERI-(around) + -OD-]
  table of the elements arranged according to a regular
  recurrence of similar properties

  *(episode, exodus, method, odometer, synod)*


## -OLE- (oil)

OLEO- (C)   [-OLE-]
  combining form meaning oil, as OLEOMETER, OLEORESIN;
  or olein, oleic, as OLEOMARGARINE
PETROLEUM (C)   [-PETR-(rock) + -OLE-]
  oily, inflammable, liquid mixture of hydrocarbons found
  naturally in certain rock strata

  *(oleaginous, oleograph, petrol)*


## -OPAC- (barrier)

OPACITY (P)   [-OPAC-]
  quality or degree of nontransparency; ability to obstruct
  radiant energy as well as light

RADIOPAQUE (P)   [-RADI-(radiant energy) + OPAQ < -OPAC-]
  impermeable to X-rays or other forms of radiant energy

  *(opaque, opaque projector)*


### -OPER- (produce an effect, work)

OPERATION (M)   [-OPER-]
  process of bringing about a change in the value or form of a quantity
OPERATOR (M)   [-OPER-]
  symbol that signifies a mathematical process

  *(cooperate, inoperative, opera, operational, operose, opus)*


### -OPT- (sight)

DIOPTER (P)   [Latin: dioptra (an ancient optical instrument for measuring heights and leveling) < DI(A)-(through) + -OPT-]
  unit for expressing the refractive power of a lens
OPTICS (P)   [-OPT-]
  branch of physics dealing with electromagnetic radiation and the phenomena of vision

  *(autopsy, biopsy, myopic, optician, optometry, stereopticon, synoptic)*


### -ORDIN- (set in order, arrange)

COORDINATE (M)   [CO-(together) + -ORDIN-]
  any of a system of magnitudes used to determine the position of a point, line, or angle by reference to a fixed system of elements; note CARTESIAN, POLAR COORDINATES

ORDER (M) [-ORD(IN)-]
   degree of complexity of an expression, equation, or operation as expressed by an ordinal number
ORDINAL (NUMBER) (M) [-ORDIN-]
   number that shows sequence or relative order of a unit in a given series
ORDINATE (M) [-ORDIN-]
   perpendicular distance of a point from the X-axis; compare ABSCISSA

   *(extraordinary, inordinate, ordain, ordinance,
   ordination, subordinate)*


## -ORGAN- (functioning part)

INORGANIC (CHEMISTRY) (C) [IN-(not) + "organic"]
   branch of chemistry dealing with elements and their compounds, with the exception of hydrocarbons and their derivatives
ORGANIC (CHEMISTRY) (C) [-ORGAN-]
   branch of chemistry dealing with the hydrocarbons and their derivatives (so named from a class of compounds which naturally exist as constituents of organized bodies, such as animals or plants)

   *(disorganize, organ, organic, organism, organization,
   reorganize)*


## -ORI- (begin, rise)

ORIENTATION (C) [Latin: sol oriens, orientis (rising sun or east: thus originally to set a map in relation to the points of a compass, then to adjust to the requirements of any situation)]
   ordering or arrangement of atoms in a compound, especially under the influence of electrical forces

89

ORIGIN (M)   [-ORI-]
> point from which measurement begins, as point of intersection of the axes of a coordinate system

*(aborigine, abortion, abortive, disorient, Orient,*
*originate, reorient)*


## -ORTHO- (straight, right)

ORTHOCENTER (M)   [-ORTHO- + "center"]
> the point of intersection of the three altitudes (perpendicular lines) of a triangle

ORTHOGONAL (M)   [-ORTHO- + -GON-(angle)]
> right-angled, mutually perpendicular; also, of variables, completely independent

*(orthochromatic, orthodontia, orthodox, orthoepy,*
*orthography, orthopedics)*


## -OSC- (swing, kiss)

OSCILLATE (P)   [Latin: oscillatus (swing) < oscillum (little mouth or face mask suspended from trees and swaying in the wind)]
> vary periodically between maximum and minimum, as an electric current; note also OSCILLOSCOPE

OSCULATE (M)   [Latin: osculatus (kiss) < osculum (little mouth, kiss)]
> touch at three or more points, as two curves

*(orotund, oscitancy, osculant, oscular, osculation)*


## -OSMOS- (push)

ENDOSMOSIS (C)   [END(O)-(within) + -OSMOS-]
> osmotic diffusion toward an inner vessel or toward the more concentrated solution; opposed to EXOSMOSIS

OSMOSIS (C)   [-OSMOS-]
  tendency of a fluid to diffuse through a semipermeable
  membrane in order to equalize concentrations on both
  sides

*(osmosis)*

**-OX(Y)-** (oxygen)

OXIDATION (C)   ["oxide" < French: oxide < OXygène +
    acIDE (acid)]
  basically the process of combining with oxygen to form an
  oxide; by extension the process of increasing the positive
  valence or decreasing the negative valence of an element
  or radical (the term *oxide* was coined originally to contrast a class of oxygen compounds which had no acid
  properties from acids (French: acide), all of which were
  supposed to contain oxygen)

OXYGEN (C)   [French: oxygène (oxygen) < -OXY-(sharp,
    acid) + -GEN-(produce)]
  element (so named because originally considered essential for all acids); note oxygen compounds DIOXIDE,
  HYDROXIDE, PEROXIDE

*(anoxia, oxyacetylene, oxymoron, paroxysm)*

**-OZ(M)-** (smell)

OZONE (C)   [-OZ-]
  unstable form of oxygen (so named from its pungent
  odor)

OSMIUM (C)   [OSM < -OZM-]
  element (so named from the pungent odor of one of its
  oxides)

*(none)*

## -PAR- (equal)

COMPARATOR (P)   [COM-(together) + -PAR-]
   measuring instrument which compares data of length, brightness, color, etc. with a fixed standard
PARITY (M, P)   [-PAR-]
   symmetrical relationship between two odd or even integers; symmetry property of a wave function

   *(comparable, disparage, disparity, impair, pair, par, peerless, umpire)*

## PARA- (beside)

PARA- (C)   [PARA-]
   prefix indicating an alternative form, especially an isomer or polymer
PARALLEL (M, P)   [PAR(A)- + -ALLEL-(one another)< -ALLO-(other)]
   of lines or planes extending in same direction and equidistant at all points; of an electrical hookup in which like poles or terminals are connected
PARAMAGNETIC (P)   [PARA- + "magnetic"]
   possessing magnetic permeability greater than that of a vacuum but less than that of a ferromagnetic substance
PARAMETER (M)   [PARA- + -METER-(measure)]
   quantity or constant whose values characterize some varying member of a system of functions or expressions

   *(parable, paradox, paramilitary, paraphernalia, paraphrase, parasite)*

## -PART- (part)

PARTICLE (P)   [Latin: particula (small part) < PARTis (part)] one of the basic components of the atom, such as proton, electron, etc.

PARTITION (M)   [Latin: partitus < PARTire (divide into parts)]
  expressing a positive number as the sum of two or more smaller positive numbers

*(compartment, impart, parse, partake, participate, partisan, partition)*

## -PATH-, -PASS- (suffer, feel)

PASSIVITY (C)   [-PASS- (suffer: thus sense of being acted upon without acting in return)]
  property of exhibiting unusual inactivity toward certain reagents

SYMPATHETIC (P)   [SYM-(together) + -PATH-(feel)]
  of sounds or vibrations caused by transmitted vibrations from a nearby vibrating body

*(antipathy, apathy, compassion, compatible, pathos, patient)*

## -PED- (foot)

PEDAL (CURVE) (M)   [-PED-]
  locus of the foot of the perpendicular let fall from a fixed point to a variable tangent to a curve

IMPEDANCE (P)   [IM < IN-(in) + -PED-(foot: thus literally entangling the feet, and thus hindering)]
  in a circuit, total opposition to flow of alternating current; in a sound-transmitting medium, ratio of force per unit area to volume displacement of a specified surface

*(expedite, impede, pedestal, pedestrian, pedigree, piedmont, sesquipedalian)*

# -PEND- (hang)

DEPENDENT (VARIABLE) (M) [DE-(down from) +
   -PEND-(hang: thus relying on)]
   variable whose value is a function of another, called
   INDEPENDENT VARIABLE

PENDULUM (P) [-PEND-]
   body hung from a fixed point and free to swing under the
   action of gravity and momentum

PERPENDICULAR (M) [Latin: perpendiculum (plumb
   line) < PER-(an intensive) + PENDere (hang)]
   line at right angles to another line or plane

SUSPENSION (C) [SUS < SUB-(up) + -PEND-(hang:
   thus hold up, support)]
   uniform diffusion of particles through a fluid

*(appendage, impend, penchant, pending, pendulous,
   penthouse, propensity)*

# -PENT(A)- (five)

PENTAGON (M) [-PENTA- + -GON-(angle)]
   polygon with five angles and five sides

PENTODE (P) [-PENT- + -OD-(path)]
   five-element electronic tube

*(pentacle, Pentagon, pentameter, Pentateuch,
   Pentecost, pentomic)*

# PER- (through; an intensive meaning "completely")

PER- (C) [PER-(an intensive)]
   prefix indicating excess amount or higher valence, as
   PERCHLORATE, PEROXIDE

94

PERMEANCE (P)  [PER-(through) + -MEA-(pass)]
 ability to be traversed by magnetic flux, reciprocal of
 RELUCTANCE

PERMUTATION (M)  [PER-(an intensive) + -MUT-
  (change)]
 process of rearranging elements in a series to achieve all
 possible changes of sequence, or any of the arrangements
 so produced

PERPENDICULAR (M)  [Latin: perpendiculum (plumb
  line) < PER-(an intensive) + PENDere (hang)]
 line at right angles to another line or plane

*(perennial, perplex, perspective, perspire, pertinacity,
 peruse, pervade)*

**PERI- (around)**

PERIGEE (P)  [PERI-(around, near) + GEE < -GEO-
  (earth)]
 nearest distance to earth of orbiting heavenly body or
 vehicle; opposed to APOGEE

PERIMETER (M)  [PERI- + -METER-(measure)]
 length of a closed curve or sum of the sides of a plane
 figure

PERIODIC (TABLE) (C)  [PERI- + -OD-(way: thus a
  going around, cycle)]
 table of the elements arranged according to a regular
 recurrence of similar properties

PERIPHERAL (P)  [PERI- + -PHER-(carry)]
 of an orbital electron in the outermost electron shell

*(periodic, periodontia, peripatetic, periphery,
 periphrasis, periscope)*

# -PETR- (rock)

PETROLEUM (C)   [-PETR- + -OLE-(oil)]
oily, inflammable, liquid mixture of hydrocarbons found
naturally in certain rock strata

SALTPETER (C)   ["salt" + -PETR-]
potassium nitrate (so named because it exudes from rocks)

*(Peter, petrel, petrify, petroglyph, petrology, petrous)*

# -PHIL- (love)

HYDROPHILIC (C)   [-HYDRO-(water) + -PHIL-]
having a high degree of affinity for water, as colloid
particles or fibres

LYOPHILIC (C)   [LYO < -LYS-(loosen, dissolve) +
-PHIL-]
characterizing a colloid system having a high degree of
attraction between the particles and the medium of dispersion

*(bibliophile, hemophilia, Philadelphia, philanthropy,*
*philosophy, philter)*

# -PHOB- (fear)

HYDROPHOBIC (C)   [-HYDRO-(water) + -PHOB-]
having little affinity for or not wet easily by water, as
colloid particles or fibres

LYOPHOBIC (C)   [LYO < -LYS-(loosen, dissolve) +
-PHOB-)]
characterizing a colloid system having little attraction
between the particles and the medium of dispersion

*(acrophobia, Anglophobe, demophobia, hydrophobia,*
*phobia, phobophobia)*

96

## -PHON- (sound)

MICROPHONE (P)  [-MICRO-(small) + -PHON-]
   device for amplifying weak sounds by converting sound
   energy into electrical energy
PHON (P)  [-PHON-]
   unit of loudness level of audible sound

   *(cacophony, euphony, phonics, phonetics,
   phonograph, symphony, xylophone)*

## -PHOR- (carry)

CHROMOPHORE (C)  [-CHROM-(color) + -PHOR-]
   grouping of atoms responsible for color in a molecule
ELECTROPHORESIS (C)  [-ELECTR-(electric) +
   -PHOR-]
   movement of charged particles in a suspension under
   influence of an electric field
PHOSPHOR (C, P)  [-PHOS-(light) + -PHOR-]
   class of substances giving off light when acted on by radiation or certain chemicals; see also PHOSPHORUS, PHOSPHOR-
   ESCENCE
PYROPHORIC (C, P)  [-PYR-(fire) + -PHOR-]
   igniting spontaneously in the presence of air

   *(anaphora, diaphoresis, metaphor, peripheral,
   periphery, semaphore)*

## -PHOS-, -PHOT- (light)

PHOSPHORESCENCE (C)  [-PHOS- + -PHOR-(carry)]
   property of continuing to glow in the dark after exposure
   to light; contrasted with FLUORESCENCE

PHOSPHORUS (C)    [Latin: Phosphorus (the Morning
    Star) < -PHOS- + -PHOR-(carry)]
    element (so named because it glows in the dark)
PHOT (P)    [-PHOT-]
    unit of illumination in the cgs system
PHOTO- (C, P)    [-PHOT-]
    combining form meaning light, as in PHOTOCONDUCTION,
    PHOTOELECTRIC, PHOTOLYSIS

*(phosphene, photic, photogenic, photograph,*
*photometer, photosynthesis)*


## -PHYS- (nature)


GEOPHYSICS (P)    [-GEO-(earth) + "physics"]
    branch of physics dealing with the earth from its inner
    core to its outer boundaries in space
PHYSICS (P)    [-PHYS-]
    study of the phenomena and laws of the physical or nat-
    ural world, especially motion, matter, and energy

*(metaphysical, neophyte, physic, physical,*
*physiognomy, physique)*


## -PIEZO- (pressure)


PIEZOELECTRICITY (P)    [-PIEZO- + "electricity"]
    electricity or electric polarity resulting from mechanical
    stresses on a crystal
PIEZOMETER (C, P)    [-PIEZO- + -METER-(measure)]
    instrument for measurement of pressure or determining
    compressibility of substances

*(none)*

## -PLAN- (flat)

COPLANAR (M)   [CO-(together) + -PLAN-]
of figures lying in the same plane

PLANE (M)   [-PLAN-]
surface such that a straight line between any two of its
points lies completely within the surface

*(airplane, esplanade, explain, pianoforte, plain,
plainsong, plan)*

## -PLANET- (wander)

PLANET (P)   [-PLANET-]
one of the heavenly bodies revolving about the sun and
shining by reflected light (so named because they seemed
to have a motion of their own among the fixed stars)

PLANETARY (P)   [-PLANET-]
of electrons found in the shells or orbits outside the
nucleus

*(planetarium, planetary transmission)*

## -PLAST-, -PLASM- (mold, shape)

PLASMA (P)   [-PLASM-]
gaseous state of matter containing nearly equal concentration of positive ions and electrons, forming a sheath at
all boundaries, and controlled with electric and magnetic
fields

PLASTICIZER (C)   [-PLAST-]
substance added to keep materials soft and flexible

*(ectoplasm, plaster, plastic, protoplasm)*

## -PLATIN- (silvery)

PLATINOID (C)  [-PLATIN-]
  silvery-white, non-tarnishing alloy of nickel, zinc, copper,
  and tungsten; also any of the six platinum metals

PLATINUM (C)  [Spanish: PLATINa (crude, native plati-
    num) < plata (silver)]
  element (so named originally from its white color, like
  silver but less bright)

*(platinum blond)*

## -PLE- (fill)

COMPLEMENTARY (M, P)  [COM-(an intensive) + -PLE-]
  of an angle or arc which will complete a right angle or
  90°; of one of a pair of spectrum colors that combine to
  form white or nearly white light

COMPLIANCE (P)  [COM-(up) + PLI < -PLE-(fill: thus
    fulfill or satisfy; in addition, association with *ply* and
    *pliant* have influenced the development of the sense of
    bending to another's will or yielding)]
  property of yielding or bending under stress

PLENUM (P)  [-PLE-]
  enclosed volume of gas under greater than surrounding
  pressure

SUPPLEMENT (M)  [SUP < SUB-(from below, thus "up
    to") + -PLE-]
  amount added to a given angle to make a straight angle
  or 180°

*(accomplish, complete, comply, depletion, implement,*
*plenitude, replenish)*

100

# -PLIC- (fold)

COMPLEX (C, M)  [COM-(together) + Latin: plectere, PLEXus (twist: thus union of several elements) < PLICare (fold)]
> of substances formed by the combining of simpler substances; of fractions having a fraction as numerator or denominator, or both; also of numbers or expressions consisting of real and imaginary numbers

EXPLICIT (RELATION) (M)  [EX-(out) + -PLIC-(fold: thus clear, direct)]
> functional relation where dependent variable is expressed directly in terms of the independent variable; opposed to IMPLICIT RELATION

MULTIPLE (M, P)  [-MULTI-(many) + PLE(X) < -PLIC-(fold: thus more than one part)]
> any of the products of some specified number and another number; group of terminals arranged for electrical connections at many points

MULTIPLICATION (M)  [-MULTI-(many) + -PLIC-]
> process of adding a number to itself a given number of times

*(accomplish, complicate, duplicity, implicate, perplex, ply, simplicity)*

## -PLUMB- (the metal lead)

PB (C)  [Latin: PlumBum (lead)]
> symbol for the element lead; note use in compounds PLUMBIC, PLUMBOUS

PLUMBAGO (C)  [-PLUMB-]
> graphite or black lead (so named because to 16th century miners plumbago mainly meant lead sulfide, but was also applied to substances similar in appearance and prop-

erties, such as graphite and molybdenum sulfide, graphite not being shown to be a form of carbon until hundreds of years later; of further interest is the derivation of lead pencil from black lead, the popular name for graphite, and of molybdenum from molybdenos, the Greek work for lead)

*(plumb, plumb line, plumbeous, plumber)*

### -POL- ( ends of an axis )

POLARIZATION (C, P)   [-POL-(ends of an axis: thus given a particular direction)]
change in the potential of an electrolytic cell opposing the direction of current; causing light waves to vibrate in a particular direction or pattern

POLE (M, P)   [-POL-]
reference point in a polar coordinate system from which a fixed line, or polar axis, is directed and from which angles are measured; either of the two points at which opposite electric or magnetic forces are concentrated, as poles of a magnet, battery, etc.

*(Polaris, polarity, polarize, Polaroid,*
*poles apart, polestar)*

### -POLA- ( calculate )

EXTRAPOLATE (M)   [EXTRA-(beyond) +
"interPOLATE"]
project beyond the range of known values

INTERPOLATE (M)   [INTER-(between) + -POLA-]
determine intermediate values between known values

*(none)*

# -POLY- (many)

POLYBASIC (C)   [-POLY- + "basic"]
  of an acid containing two or more hydrogen atoms which
  can be replaced by basic atoms or radicals

POLYGON (M)   [-POLY- + -GON-(angle)]
  closed plane figure bounded by straight lines or arcs,
  usually more than four

POLYMERIZATION (C)   [-POLY- + -MER-(part)]
  reaction which combines smaller molecules to form larger
  molecules with repeating structural units and same percentage composition

POLYNOMIAL (M)   [-POLY- + -NOMI-(name)]
  algebraic expression containing two or more terms

*(hoi polloi, polyandry, polygamy, polyglot, polyp,
polytechnic, polytheism)*

# -PON-, -POS- (place)

COMPONENT (C, P)   [COM-(together) + -PON-]
  ingredient of a mixture; one of the elements making up a
  vector quantity

COMPOUND (C)   [COM-(together) + -PO(U)N(D)-]
  chemically distinct substance resulting from the combining of elements or radicals in fixed proportions; distinguished from MIXTURE

EXPONENT (M)   [EX-(out) + -PON-(place: thus literally
    placed out from a quantity)]
  number or quantity placed above and to the right of an
  expression to signify an operation to be performed

TRANSPOSE (M)   [TRANS-(across) + -POS-]
  move a term with changed sign to the other side of the
  equation

*(apposite, depose, exposition, juxtapose,
opponent, postpone, propound)*

**103**

## PRE- (before)

PRECESSION (OF EQUINOXES) (P)  [PRE- + -CESS-
(go)]
   earlier occurrence of equinoxes in each sidereal year
   caused by slow changes in direction of earth's axis
PRECIPITATE (C)  [PRE- + -CIPIT- < Latin: caput
(head: thus headfirst, falling)]
   become insoluble, separate, and settle

*(precedent, precipice, preclude, precocious,
prejudice, presentiment)*


## -PRESS- (press)

COMPRESSIBILITY (P)  [COM-(together) + -PRESS-]
   coefficient indicating volume change of a substance per
   unit pressure
PRESSURE (P)  [-PRESS-]
   force per unit area exerted over a surface

*(depress, express, impress, oppress, print, repress,
suppress)*


## -PRIM- (first)

PRIMARY (C, P)  [-PRIM-]
   designating an initial replacement of one atom or molecule, or a carbon atom united to only one other in a molecule; of a battery cell whose electrochemical action cannot be reversed and thus recharged, or of the basic colors
   from which all others are derived
PRIME (M)  [-PRIM-]
   of a whole number divisible only by itself and unity

*(prima donna, prima-facie, primate, primeval,
primitive, primogeniture)*

# PRO- (forward)

PROGRESSION (M)   [PRO- + -GRESS-(go)]
   series of numbers or quantities, each derived from pre-
   ceding by a constant principle
PROMOTER (C)   [PRO- + -MOT-(move)]
   substance which in small amounts will speed up the action
   of a catalyst
PROTRACTOR (M)   [PRO- + -TRACT-(draw)]
   instrument for measuring and laying off angles
RECIPROCAL (M)   [RE-(backward) + PRO-]
   involving a mutual relation between quantities; quotient
   resulting from the division of unity by a number or
   expression

*(procrastinate, progenitor, prognosis, prologue,*
*promontory, prosecute)*

# -PROP(E)R- (belonging to)

PROPER (M)   [-PROPER-]
   of fractions whose value is less than unity (so named from
   the sense of strict, literal, to which the name accurately
   belongs); opposed to IMPROPER
PROPERTY (C, P)   [-PROPER-]
   characteristic quality of a substance or body belonging to
   all members of that class

*(appropriation, expropriate, impropriate,*
*inappropriate, propriety)*

# -PROT(O)- (first)

PROT(O)- (C)   [-PROT(O)-]
   combining form indicating the first or lowest member of
   a series having the lowest proportion of the element or
   radical specified, as PROTOCHLORIDE, PROTOXIDE

PROTON (P)   [-PROT-]
   one of the fundamental particles in the nucleus of the
   atom

   *(protagonist, protein, protocol, protomartyr,
   protoplasm, prototype)*


### -PROX- (nearest)

PROXIMATE (ANALYSIS) (C)   [-PROX-]
   of a quantitative analysis determining the percentage of
   components directly making up a substance, such as
   moisture, volatile matter, etc. in a sample of coal (so
   named because these are first determined in an analysis);
   compare ULTIMATE ANALYSIS
PROXIMITY (EFFECT) (P)   [-PROX-]
   mutual effect of current variations in neighboring conductors

   *(approximate, propinquity, proximate, proximity)*


### -PSEUDO- (resembling, false)

PSEUDO- (C)   [-PSEUDO-]
   combining form meaning resembling, related, as PSEUDO-
   CRYSTALLINE, PSEUDOINSOLUBLE, PSEUDOSALT
PSEUDOSCIENCE (S)   [-PSEUDO- + "science"]
   system or doctrine erroneously considered to be scientific, as ASTROLOGY, NUMEROLOGY

   *(pseudepigrapha, pseudoclassic, pseudonym)*


### -PULS- (push)

IMPULSE (P)   [IM < IN-(on) + -PULS-]
   force acting for a relatively brief period; change in momentum because of a force

PULSE (P)   [-PULS-]
 transient surge of electrical or electromagnetic energy

 *(compulsion, dispel, expulsion, impulsive, propulsion,*
 *pulsate, repulsion)*

## -PUNCT- (point)

COPUNCTAL (M)   [CO-(together) + -PUNCT-]
 meeting in a point or having a point in common
PUNGENT (C)   [PUNG < -PUNCT-(point: thus sharp)]
 having a sharp, acrid taste or smell

 *(compunction, expunge, point, punctilious,*
 *punctual, punctuate, puncture)*

## -PYR (O)- (fire)

PYRITES. (C)   [Greek: pyrites lithos (fire stone; originally
 a flint or stone used for striking fire)]
 any of the metallic sulfides, especially PYRITE (iron pyrites
 or fool's gold)
PYRO- (C, P)   [-PYRO-]
 combining form meaning heat, as PYROCONDUCTIVITY,
 PYROMETER
 *(empyrean, pyracantha, pyre, pyretic, pyrotechnics)*

## -QUADR(I)- (four, square)

QUADRATURE (M)   [-QUADR-]
 any of the four parts formed by the intersection of the
 X- and Y-axes, or a quarter section of a circle
QUADRATIC (M)   [-QUADR-]
 of an equation containing the square but no higher power
 of the unknown

QUADRATURE (M) [-QUADR-]
    determination of the dimensions of a square equal in area
    to a given surface, usually one bounded by a curve
QUADRILATERAL (M)   [-QUADRI- + -LATER-(side)]
    four-sided polygon

*(quadrangle, quadrille, quadruped, quadruple, quadruplet,*
*quatrain, quartet)*

## -QUAL- (what kind)

QUALITATIVE (ANALYSIS) (C)   [-QUAL-]
    determining only the kind and number of constituents in a
    substance; distinguished from QUANTITATIVE ANALYSIS
QUALITY (P)   [-QUAL-]
    in acoustics, the property of a tone that distinguishes it
    from another having same pitch and intensity but differ-
    ent overtones; note also the term Q (from initial letter of
    "quality factor"), a measure of the efficiency of tuned
    circuits

*(disqualify, qualification, quality)*

## -QUANT- (how much)

ALIQUANT (M)   [-ALI-(other) + -QUANT-]
    not dividing evenly into another number but leaving a
    remainder; opposed to ALIQUOT
QUANTITATIVE (ANALYSIS) (C)   [-QUANT-]
    determining the amounts or percentages of constituents
    in a substance; distinguished from QUALITATIVE ANALYSIS
QUANTITY (M)   [-QUANT-]
    expression of value or magnitude
QUANTUM (P)   [-QUANT-]
    elementary unit or discrete particle of energy emitted
    from radiating bodies

*(quantify, quantitative)*

## -QUOT- (how many)

ALIQUOT (C, M)  [-ALI-(other) + -QUOT-]
    measured proportion of the volume of a solution; dividing
    into another number without remainder; opposed to
    ALIQUANT

QUOTIENT (M)  [-QUOT-]
    number signifying how many times a divisor will go into
    a dividend

*(quota, quotidian)*

## -RADI- (ray)

RADIAN (M)  ["RADIus" < -RaDI-]
    angle subtended at the center of a circle by an arc equal
    in length to the radius

RADIATION (C, P)  [-RADI-]
    emission of radiant energy as waves or particles (so
    named from the ray-like emission from a center); note
    also chemical elements RADIUM, RADON

RADIO- (C, P)  [-RADI-]
    combining form meaning radiant energy, as RADIOACTIVE,
    RADIOCARBON; or radio, as RADIOSONDE

RADIUS (M)  [-RADI-]
    straight line drawn from center to outer boundary of circle
    or sphere

*(irradiate, radial, radiant, radiator, ray)*

## -RADIC- (root)

RADICAL (C, M)  [-RADIC-]
    group of atoms forming the base of a compound and

remaining unchanged during ordinary reactions; quantity forming or expressed as the root of another (note that symbol $\sqrt{\phantom{xx}}$ is a modification of *r* in Latin radix—root)

RADIX (M)   [-RADIC-]
number made the base of a scale of enumeration

*(eradicate, radicalism, radicel, radish)*

#### -RAR- ( scarce )

RARE (EARTHS) (C)   [-RAR-]
oxides of the rare-earth metals found in minerals which are widely distributed but relatively scarce
RAREFACTION (C, P)   [-RAR- + -FACT-(make)]
process of making gases less dense; region of minimum pressure in a compression wave medium

*(rara avis, rarefy, raree show, rarity)*

#### -RATIO- ( reason )

RATIO (M)   [-RATIO-(reason: thus a fixed relationship)]
relation between two similar quantities, determined by the number of times one contains the other
RATIONAL (M)   [-RATIO-(reason: thus able to be worked with)]
of a number or function which can be expressed without the use of the radical sign as an integer or quotient of integers, opposed to IRRATIONAL; note also RATIONALIZE, remove radical signs from an equation to make it more workable

*(irrationality, rate, ratify, ratiocination, ration, rationale)*

## RE- (back, again)

REACTION (C, P)   [RE- + "act"]
 change involving action of substances upon each other; force acting in opposition to a given force

REFRACTION (P)   [RE- + -FRACT-(break)]
 bending of a light ray from a straight path in passing through mediums of different density

REGELATION (P)   [RE- + -GEL-(congeal)]
 process of melting and freezing again when pressure is removed

RELUCTANCE (P)   [RE- + -LUCT-(fight)]
 opposition to passage of magnetic lines of force; reciprocal of PERMEANCE

*(recluse, redundant, regress, remorse, renaissance, residue, resilient)*

## -RECT- (guide, straight)

CORRECT (P)   [COR < COM-(together) + -RECT-(straight)]
 cause to conform to a standard, as an instrument reading, or a lens to be free of aberration

DIRECTRIX (M)   [DI-(apart from) + -RECT-(guide: thus to control, direct)]
 fixed line that determines the motion of another line generating a curve or surface

RECTIFY (C, M, P)   [-RECT-(straight) + -FY-(make)]
 repeatedly distill a liquid for the purpose of purification; ascertain the length of a curve; convert alternating into direct current

REGULAR (M)   [REG < -RECT-(guided: thus according to rule)] of a figure with equal sides and angles, or an equation controlled by a single law or operation throughout

*(corrigible, directive, erect, rectitude, rector, regent, regulate, regimen)*

## -RHEO- (flow)

RHEOLOGY (P)   [-RHEO- + -LOGY-(study of)]
study of the deformation and flow of matter
RHEOSTAT (P)   [-RHEO- + -STAT-(standing still)]
variable resistor for regulating the flow of electric current

*(catarrh, diarrhea, hemorrhoid, logorrhea, rheum, rheumatism)*

## -RHOD- (rose)

RHODIUM (C)   [-RHOD-]
element (so named from its generally rose-colored salts)
RHODIZITE (C)   [-RHOD-]
borate of aluminum and potassium (so named because it colors the blowpipe flame red)

*(rhododendron, rhodolite)*

## -RIV- (stream)

DERIVATION (M)   [DE-(from) + -RIV-(stream: thus drawing off water, obtaining from a source)]
process of obtaining one function from another according to some definite principle
DERIVATIVE (C, M)   [DE-(from) + -RIV-]
compound obtained from another, usually by partial replacement; instantaneous rate of change of a function referred to a variable

*(derive, rival, rivulet)*

112

## -ROS- (gnaw)

CORROSION (C)   [COR < COM-(an intensive) + -ROS-]
   slow eating into or wearing away of metal surfaces by
   chemical action
EROSION (C)   [E-(out, off) + -ROS-]
   gradual wearing away by friction or eating into by chemical action

*(corrosive, erode, rodent, rodenticide)*

## -ROTA- (turn, wheel)

ROTOR (P)   ["ROTatOR" < -ROTA-]
   revolving part of an alternating-current motor; distinguished from STATOR
DEXTROROTATORY (C)   [-DEXTR-(right) + -ROTA-]
   of a substance which turns the plane of polarization of
   light to the right or clockwise; opposed to LEVOROTATORY

*(rodeo, rotary, rotunda, rotate, rotogravure, roué,
roulette)*

## -SATUR- (full)

SATURATION (C, P)   [-SATUR-]
   condition of a solution containing the maximum dissolved
   amount, or of a substance combined to maximum extent
   of its combining capacity; degree of purity of a color as
   measured by its freedom from dilution with white, or the
   highest possible degree of magnetization
SUPERSATURATE (C)   [SUPER-(beyond) + "saturate"<
   -SATUR-]
   saturate a solution beyond the normal point

*(assets, insatiable, sate, satiate, satisfy)*

## -SCAL-, -SCEND- (climb, ladder)

SCALAR (M)   [-SCAL-(ladder)]
  of a quantity that has magnitude only, as measured by a
  point on a line or scale; distinguished from VECTOR
TRANSCENDENTAL (M)   [TRANS-(beyond) + -SCEND-
    (climb)]
  of numbers and functions that cannot be produced by a
  finite number of the ordinary operations of algebra

*(ascend, condescend, descend, escalate, escalator,*
*scale, scansion)*


## -SCOP- (observe)

-SCOPE (C, P)   [-SCOP-]
  combining form meaning an instrument designed for ob-
  serving, as OSCILLOSCOPE, POLARISCOPE
-SCOPY (C, P)   [-SCOP-]
  combining form meaning observation, examination, as in
  CRYOSCOPY, MICROSPECTROSCOPY

*(bishop, episcopal, kaleidoscopic, microscopic,*
*scope, stethoscope)*


## -SCRIB-, -SCRIPT- (write)

CIRCUMSCRIBE (M)   [CIRCUM-(around) + -SCRIB-]
  draw a figure around another with maximum number of
  contact points
DESCRIPTIVE (GEOMETRY) (M)   [DE-(down) +
    -SCRIPT-]
  geometric system using plane projections to solve spatial
  problems

114

ESCRIBE (M)  [E-(out of) + -SCRIB-]

  draw a circle so as to touch one side of a triangle and the prolongation of the other two sides

INSCRIBE (M)  [IN-(in) + -SCRIB-]

  draw a figure inside another with maximum number of contact points

*(conscript, postscript, prescribe, script, Scripture, subscribe, transcribe)*

## -SEC-, -SEQU- (follow)

CONSEQUENT (M)  [CON-(together) + -SEQU-]
  second term of a ratio

SECOND (M)  [Latin: pars minuta secunda (second small part: i.e., the result of the second operation of sexagesimal division, the first being the prime minute, now simply "minute") < SECunda (following, second)]
  unit of angular measure, 1/60 of a minute of arc

SECONDARY (P)  [-SEC-]
  of a cell that can be recharged by passage of current in the opposite direction; of the emission of electrons from a substance excited by electrons or ions from a primary source; of current produced by induction

SEQUENCE (M)  [-SEQU-]
  ordered succession of terms

*(consecutive, obsequious, persecute, prosecute, sect, sequel, subsequent)*

## -SEC(T)- (cut)

BISECT (M)  [BI-(double)- + -SECT-]
  divide into two equal parts

INTERSECTION (M)  [INTER-(between) + -SECT-]
  point or line common to two or more lines, planes, or surfaces

SECANT (M)   [-SEC-]
in geometry, a straight line cutting a given curve in two or more parts; in trigonometry, originally the length of a straight line drawn from the center of a circle through the end of the arc of its circumference and terminated by a line tangent to the radius at the other end of the arc: in modern use, the ratio of this line to the radius, or as the function of an angle, the ratio of the hypotenuse of a right-angled triangle to that of one side of the included angle

SECTION (M)   [-SECT-]
figure formed by the intersection of a solid and a plane

(dissect, insect, resection, scion, sector, segment,
sickle, transect)


## -SED- (sit)

RESIDUAL (C, M)   [RE-(back) + SID < -SED- (sit: thus remain behind)]
of the product or substance left over at the end of a chemical process (note also RESIDUE); of the difference between observed results and computation from a formula, or of the difference between the mean of a series of values and any one of them

SEDIMENT (C)   [-SED-]
insoluble matter settled or precipitated from a suspension

(assiduous, dissident, preside, reside, sedate, sedentary,
supersede)


## -SELEN- (moon)

SELENITE (C)   [Greek: selenites lithos (moon stone)]
variety of gypsum (so named because it was anciently thought to wax and wane with the moon)

SELENIUM (C)   [-SELEN-]
> chemical element (so named because of its association with the similar element TELLURIUM, named after the earth)

*(Selene, selenography)*

## SEMI- (partial, half)

SEMICONDUCTOR (P)   [SEMI- + "conductor" < CON- (together) + -DUCT-(to lead)]
> substance whose conductivity is between that of metals and non-metals

SEMIPERMEABLE (C)   [SEMI- + "permeable" < PER- (through) + -MEA-(pass)]
> permitting the passage of smaller molecules but not larger ones

*(semicircle, semicolon, semifinal, semipostal,*
*semiprecious, semivowel)*

## -SENS- (feel)

SENSITIVITY (P)   [-SENS-]
> degree of responsiveness of an instrument or radio receiver to electric current or radio waves

SENSOR (P)   [-SENS-]
> instrument or device which detects and responds to a stimulus or signal

*(assent, consensus, dissension, presentiment,*
*resent, sense, sensual)*

## -SEQU- (*see* -SEC-)

117

## -SICC- (dry)

DESICCATOR (C)  [DE-(an intensive) + -SICC-]
    device for keeping residues or specimens free from mois-
    ture by the use of a DESICCANT
SICCATIVE (C)  [-SICC-]
    substance which has drying properties

*(desiccate, exsiccate)*

## -SIDER- (star)

SIDEREAL (P)  [-SIDER-]
    measured by the stars, as SIDEREAL YEAR, period of time in
    which the sun makes a complete revolution with reference
    to the fixed stars
SIDEROSTAT (P)  [-SIDER- + -STAT-(stand)]
    turning mirror which constantly reflects the light of a star
    into a fixed telescope

*(consider, desire, desideratum)*

## -SIL- (*see* -SULT-)

## -SILIC- (flint)

SILICON (C)  [Latin: SILICis (flint)]
    element (so named because it occurs abundantly as silica,
    or silicon dioxide, the principal constituent of quartz; the
    spelling was changed from earlier silicium to conform to
    the pattern of the non-metals, boron and carbon)
SILICONES (C)  ["silicon"]
    plastic-like compounds containing a silicon-carbon bond

*(silex, silicify, silicosis)*

## -SIMIL- ( like )

SIMILAR (M)   [-SIMIL-]
  having the same shape but different size and position
SIMULTANEOUS (EQUATIONS) (M)   [Latin: simul (at
    same time) < SIMILis (like)]
  group of equations satisfied by the same values of the
  unknowns

*(assimilate, dissimilar, dissimulate, simile,*
*similitude, simulate)*

## -SIST- ( *see* -STA- )

## -SOCI- ( join )

ASSOCIATIVE (M)   [AS < AD-(to) + -SOCI-(join: thus
    combine, unite)]
  of an operation that can be applied to a group of terms
  or factors in any order
DISSOCIATION (C)   [DIS-(apart) + -SOCI-]
  breakdown of a compound into simpler constituents, or of
  an electrolyte into ions

*(asocial, association, soccer, sociable, socialize,*
*society, sociology)*

## -SOL- ( sun )

INSOLATION (P)   [IN-(in) + -SOL-]
  rate of solar radiation per unit area
SOLSTICE (P)   [-SOL- + -STIT-(stand)]
  time of year when sun is furthest from celestial equator
  (so named because sun seems to pause before returning)

*(parasol, solano, solar, solarium, solar plexus)*

119

## -SOLID- (firm)

SOLID (M, P)   [-SOLID-]
> three-dimensional figure; state of matter having definite shape and volume

SOLIDUS (C)   [-SOLID-]
> line in a phase diagram indicating the upper limit of temperature for the solid phase

> *(consolidate, solder, soldier, solidarity, solidify)*

## -SOLV-, -SOLUT- (loosen)

ABSOLUTE (C, M, P)   [AB-(from) + -SOLUT-]
> free from imperfection, qualification, dependence: as AB-SOLUTE ALCOHOL, free from water; ABSOLUTE CONSTANT, unchanging in value; ABSOLUTE DENSITY, actual, not relative; ABSOLUTE (CGS) SYSTEM OF UNITS, using fundamental units of space, mass, and time, as ABAMPERE, ABOHM; ABSOLUTE VALUE, without regard to sign; ABSOLUTE ZERO, free from molecular movement

RESOLVE (M, P)   [RE-(back) + -SOLV-(loosen: thus to separate into parts)]
> carry out required operations and solve an equation; separate spectral lines or vector forces into their components, or cause the individual parts of a lens image to be distinguished

SOL (C)   ["SOLution"]
> colloidal suspension in a liquid: in water, HYDROSOL, in a gas, AEROSOL

SOLUTION (C, M)   [-SOLUT-]
> homogeneous system formed by mixing or DISSOLVING (DIS- "apart") a solid, liquid, or gaseous substance (SOLUTE) with a liquid (SOLVENT); required answer to a problem or method of obtaining it

> *(absolution, absolve, dissolute, insolvent, irresolute, resolution, solve)*

# -SON- ( sound )

CONSONANCE (P)   [CON-(together) + -SON-]
acoustic or electrical resonance between independent systems or bodies

DISSONANCE (P)   [DIS-apart) + -SON-]
condition of harmonically unresolved sounds that produce beats

RESONANCE (C, P)   [RE-(again) + -SON-(sound: thus to echo)]
property of molecules having two or more electronic structures differing only in electron positions; mechanical or electrical vibrations of relatively large amplitude caused by a relatively small periodic stimulus of approximately the same period

SONIC (P)   [-SON-]
of the frequency of audible sound; of sound waves used for work other than communication; note compounds SUB-SONIC, TRANSONIC, SUPERSONIC, HYPERSONIC

*(assonance, consonant, resonant, sonar, sonata, sonorous, unison)*

# -SORPT- ( suck up )

ABSORPTION (C, P)   [AB-(from) + -SORPT-]
process of taking up and retaining internally a liquid or gas by a porous substance; process of taking in and transforming radiant energy into a different form

ADSORPTION (C)   [AD-(to) + -SORPT-]
taking up a substance at the solid surface of another; removal is called DESORPTION (DE- "reversal")

*(absorb, absorbed, absorbent, sorbefacient)*

121

# -SPECT- (look)

PERSPECTIVE (S)   [PER-(through) + -SPECT-(look:
     originally looking through an optical instrument, then
     appearance, view)]
   technique of representing an object on a flat surface as it
   appears to the eye
SPECTRO- (P)   [-SPECT-]
   combining form meaning radiant energy as displayed in a
   spectrum, as SPECTROGRAPH, SPECTROSCOPE
SPECTRUM (P)   [-SPECT-(look: thus an image)]
   series of colored images into which a beam of white light
   is separated by a prism or diffraction grating; any array of
   of wavelength frequencies resulting from dispersion of
   radiant energy
SPECULAR (P)   [Latin: SPECULum (mirror) < SPECTus
     (look)]
   of reflection from an extremely smooth, mirrorlike surface

*(aspect, circumspect, conspicuous, inspect,
    prospect, retrospect, specious)*


## -SPERS- (scatter)

DISPERSION (C, M, P)   [DI(S)-(away) + -SPERS-]
   scattering and suspension of fine particles (DISPERSED
   PHASE) in a liquid medium (DISPERSION MEDIUM); scattering from their average of a series of values in a frequency distribution; separation of light into different
   colors by refraction

*(asperges, aspergill, aspersion, disperse,
    intersperse, sparse, sparsity)*

122

# -SPHER- (ball)

SPHERE (M)  [-SPHER-]
   figure whose surface is equidistant from the center at all
   points; note also HEMISPHERE (HEMI-"half")
-SPHERE (P)  [-SPHER-]
   combining form meaning enveloping like a sphere, as
   ATMOSPHERE, GRAVISPHERE, IONOSPHERE

   *(atmosphere, ensphere, planisphere, spherical)*


# -STA-, -STIT-, -SIST- (stand)

RESISTANCE (P)  [RE-(back) + -SIST-(stand: thus to
      withstand)]
   opposition to the passage of a current by a conductor; reciprocal of CONDUCTANCE
STABLE (C, P)  [-STA-]
   of compounds that resist decomposition; of a condition
   resistant to forces disturbing equilibrium, or of nonradio-
   active bodies
STATICS (P)  [-STA-]
   branch of dynamics dealing with bodies at rest or in
   equilibrium; note compounds ELECTROSTATICS, HYDROSTAT-
   ICS, ISOSTATIC
SUBSTITUTION (C, M)  [SUB-(in place of) + -STIT-
      (stand, put)]
   replacement of an atom or molecule by another (called
   SUBSTITUENT) in a reaction; method of replacing one expression for another of different composition but equal
   value

   *(consist, distant, establish, exist, interstice,
   obstacle, persist, station)*

## -STELLA- (star)

CONSTELLATION (P)  [CON-(together) + -STELLA-]
  group of stars forming an outline of a picture and so
  named
STELLARATOR (P)  ["stellar" < -STELLA-]
  device for heating plasma to thermonuclear temperatures
  (so named because of the high temperatures developed)

*(stellar, stellate, stelliform, stellular)*

## -STER- (solid)

STERADIAN (M)  [-STER- + "radian"]
  unit of measurement of solid angles
STEREO- (C, P)  [-STER-]
  combining form meaning of spatial arrangement, three-
  dimensional, as STEREOCHEMISTRY, STEREOISOMERISM,
  STEREOPHONIC

*(stere, stereobate, stereopticon, stereoscope, stereotype)*

## -STIT- (*see* -STA-)

## -STRING-, -STRICT- (draw tight)

STRAIN (P)  [French: estraindre (wring hard, strain) <
  Latin: STRINGere (draw tight)]
  deformation in a solid body due to STRESS [French: es-
  trecier (pinch) < Latin: STRICTUS (draw tight)]
-STRICTION (P)  [-STRICT-]
  combining form meaning deformation, as ELECTROSTRIC-
  TION, MAGNETOSTRICTION

*(astringent, constraint, distress, restrain,*
*restrict, stringent, strict)*

124

# -STRUCT- (build up)

CONSTRUCT (M)   [CON-(together) + -STRUCT-]
: draw a figure to satisfy certain conditions

STRUCTURE (C)   [-STRUCT-]
: spatial arrangement of atoms in a molecule or molecules in a compound

> *(constructive, construe, destruct, instruct,*
> *obstruction, superstructure)*

# SUB- (under)

SUB- (C, P)   [SUB-]
: prefix indicating less than: as less than normal amount (SUBOXIDE), of particles of less than atomic mass (SUB-ATOMIC), of frequencies below human audibility (SUB-SONIC)

SUBLIMATE (C)   [SUB-(from below, thus "up to") + -LIM-(lintel: thus to raise up)]
: change from solid to gaseous state and back to solid without apparent liquefaction

SUBTEND (M)   [SUB- + -TEND-(stretch)]
: extend under, and thus be opposite to and mark off

SUBTRACTION (M)   [SUB-(from below, away) + -TRACT-(draw)]
: operation of deducting one quantity from another; note also SUBTRAHEND, number to be subtracted from another

> *(sublime, submerge, subsidy, subsume, subversion,*
> *suppress, suspend)*

# -SULT-, -SIL- (leap)

RESILIENCE (P)   [RE-(back) + -SIL-]
: ability of an elastically strained body to return to its original size and shape after deformation

RESULTANT (P) [RE-(back) + -SULT-(leap: thus to rebound, follow from as a consequence)]
   force or velocity equivalent in effect to two or more such forces acting together

   *(assail, desultory, exult, insult, result, salient, sally, somersault)*

### SUPER- (above, higher)

SUPERNATANT (C) [SUPER- + -NAT-(swim)]
   of the liquid lying above a precipitate
SUPERSONIC (P) [SUPER- + "sonic" < -SON-(sound)]
   of speeds above speed of sound, from M 1.8 to M 5

   *(superb, superficial, superfluous, superlative, supersede, surpass)*

### SYN-, SYM- (together)

SYMMETRICAL (C, M) [SYM- + -METR-(measure)
   showing a definite repeated arrangement of atoms in the structural formula; of an equation, expression, or relation whose terms are interchangeable without affecting its validity
SYNCHRONOUS (P) [SYN- + -CHRON-(time)]
   having the same period and phase
SYNCHROTRON (P) [SYN- + -CHRO(N)-(time) + -TRON (device)]
   apparatus for accelerating atomic particles (so named because the accelerating forces synchronize with the movement of the particles)
SYNTHESIS (C) [SYN- + -THESIS-(place)]
   producing a complex compound by combining simpler compounds, radicals, or elements; compare ANALYSIS

   *(symbol, sympathy, symptom, synagogue, sync, synonym, synopsis, syzygy)*

## -TAB(U)L- (arrangement, list)

TABLE (C, M, P)   [-TABL-]
   arrangement of data which exhibits facts or relations in an
   orderly and convenient way
TABULAR (DIFFERENCE) (M)   [-TABUL-]
   varying difference between successive values of a function

*(entablature, retable, tableau, table d'hôte,*
*tablet, tabulate, tabulator)*

## -TACH- (quick)

TACHOMETER (P)   [-TACH- + -METER-(measure)]
   instrument for measuring speed of rotation or rate of flow
   in liquids
TACHYLITE (C)   [-TACH- + -LYT-(loosen)]
   variety of basaltic glass (so named because it rapidly
   decomposes in acids)

*(tachina fly, tachistoscope, tachycardia, tachygraphy)*

## -TANG-, -TACT- (touch)

CONTACT (M, P)   [CON-(together) + -TACT-]
   touching of curves or surfaces resulting in common tan-
   gents; point of junction of two conductors in a circuit
INTEGER (M)   [IN-(not) + TEG < -TANG-(touch: thus
   untouched, whole)]
   whole number
SUBTANGENT (M)   [SUB-(under) + "tangent"]
   that part of the axis of a curve contained between the
   tangent to a given point and the ordinate of the point
TANGENT (M)   [-TANG-]
   line or surface touching and not intersecting a curve or
   surface at a point; also trigonometric function or ratio

(originally "tangent line," length of a straight line perpendicular to the radius touching one end of the arc and terminated by the secant line)

*(attain, contagious, contiguous, contingent,*
*intact, tact, tangible)*

### -TAUTO- (the same)

TAUTOCHRONE (P)   [-TAUTO- + -CHRON-(time)]
  inverted cycloid representing force of gravity: time of descent from every point to lowest point is the same
TAUTOMERISM (C)   [-TAUTO- + -MER-(part)]
  property of existing in equilibrium between two isomeric forms and of reacting to form either

*(tautology)*

### -TECHN- (practical skill)

TECHNETIUM (C)   [-TECHN-]
  element (so named because it was artificially produced)
TECHNOLOGY (S)   [-TECHN-]
  applied or practical science

*(architect, pyrotechnics, technician, technique,*
*technocracy, tectonic)*

### -TEKT- (melt)

EUTECTIC (C)   [EU-(good) + TECT < -TEKT-]
  mixture which has lowest possible melting point
TEKTITES (P)   [-TEKT-]
  rounded glasslike objects of varied weight and color generally believed to have originated in interplanetary space

*(none)*

## -TELE- (at a distance)

TELEMETRY (P)   [-TELE- + -METR-(measure)]
> technique of recording and transmitting data from a distance, as with missiles and rockets.

TELESCOPE (P)   [-TELE- + -SCOP-(observe)]
> instrument which enlarges distant objects

> *(telecast, telegnosis, telepathy, telephoto,*
> *teletype)*


## -TEMPER- (mix in proper proportions)

TEMPER (C)   [-TEMPER-]
> bring to desired condition by mixing or treating

TEMPERATURE (C, P)   [-TEMPER-]
> measured degree of hotness or coldness (originally a temperate or moderate condition of climate, then climatic condition in regard to heat or cold, and finally measured degree of hotness or coldness in anything)

> *(attemper, distemper, intemperate, tempera,*
> *temperament, temperance)*


## -TEND-, -TENS- (stretch)

EXTENSION (P)   [EX-(out) + -TENS-]
> property of matter which enables it to occupy space

INTENSITY (P)   [IN-(in, at) + -TENS-(stretch: thus stretched out, to an extreme degree)]
> measured force or energy of heat, light, radiation, electric current, etc. per unit quantity

SUBTEND (M)   [SUB-(under) + -TEND-]
> extend under, and thus be opposite to and mark off; note also HYPOTENUSE (HYPO-"under") side under or subtended by the right angle

**TENSILE (STRENGTH) (P)  [-TENS-]**
 resistance of a material to lengthwise stress or rupture

*(distend, intend, ostensible, portend, pretend,
tendon, tension, tenterhook)*


**-TEN(T)- (*see* -TIN-)**


**-TERM- (limit)**


**DETERMINATE (M)  [DE-(an intensive) + -TERM-]**
 having a fixed value or limited number of solutions
**TERMINAL (P)  [-TERM-]**
 connection point to an electrical apparatus or circuit; of
 the velocity of a falling body when resistance of air equals
 force of gravity

*(conterminous, determine, exterminate, interminable,
termination, terminus)*


**-TETR(A)- (four)**


**TETR(A)- (C)  [-TETR(A)-]**
 combining form indicating four atoms or radicals, as
 TETRACHLORIDE, TETRACID, TETROXIDE
**TETRAD (P)  [-TETRA-]**
 atom, radical or element which has a valence of four

*(Tetragrammaton, tetralogy, tetrameter, tetrapod, tetrarch)*


**-THERM- (heat)**


**DIATHERMANCY (P)  [DIA-(through) + -THERM-]**
 property of being capable of transmitting infrared or heat
 rays

130

ENDOTHERMIC (C)   [ENDO-(within) + -THERM-]
 of a change accompanied by or produced from heat absorption; opposed to EXOTHERMIC
ISOTHERMAL (P)   [-ISO-(equal) + -THERM-]
 of a relation between variables of volume or pressure under conditions of uniform temperature
THERMO- (C, P)   [-THERM-]
 combining form meaning heat, as in THERMOCHEMISTRY, THERMOCOUPLE, THERMODYNAMICS, THERMOELECTRIC

 *(diathermy, thermal, thermometer, thermonuclear, thermos, thermostat)*


## -THESIS- (place)

APOTHEM (M)   [APO-(from) + THEM < -THESIS-]
 perpendicular drawn from the center of a regular polygon to any side
HOMOTHETIC (M)   [-HOMO-(same) + THET < -THESIS-]
 of similar and similarly placed geometric figures
METATHESIS (C)   [META-(across) + -THESIS-(place: thus interchange)]
 reaction involving interchange of elements or radicals between compounds: double decomposition
SYNTHESIS (C)   [SYN-(together) + -THESIS-]
 producing a complex compound by combining simpler compounds, radicals, or elements; compare ANALYSIS

 *(anathema, antithesis, epithet, hypothesis, parenthesis, prosthesis, thesis)*


## -TIN-, -TEN(T)- (hold)

CONTAIN (M)   [CON-(together, in) + -T(A)IN-(hold: thus include within itself)]
 be a multiple of, or divisible by without remainder

CONTINUOUS (PHASE) (C)   [CON-(together) + -TIN-]
   another name for the dispersion medium, in which the
   particles of a colloid are suspended

CONTINUUM (M, P)   [CON-(together) + -TIN-(hold:
      thus to be continuously connected)]
   infinite set of numbers or points between any two of which
   a third may be interpolated; continuous magnitude or
   extent, as the four-dimensional SPACE-TIME CONTINUUM

RETENTIVITY (P)   [RE-(back) + -TENT-]
   capacity for retaining magnetism after the magnetizing
   force has been removed

*(abstain, continent, pertinent, sustain, tenable,*
*tenacity, tenet, tenure)*


### -TOM- (cut)

ATOM (C)   [A-(not) + -TOM-]
   smallest particle of an element that can exist alone and in
   combination and that can not be changed or destroyed by
   chemical means (so named because originally considered
   an indivisible particle)

MICROTOME (C)   [-MICRO-(very small) + -TOM-]
   instrument for cutting extremely thin sections of speci-
   mens

*(anatomy, appendectomy, dichotomy, entomology,*
*epitome, tome)*


### -TON- (stretching, tension)

HYPERTONIC (C)   [HYPER-(above) + -TON-]
   of a solution having greater osmotic pressure than another;
   compare HYPOTONIC (HYPO-"below, less")

ISOTONIC (C)   [-ISO-(equal) + -TON-]
   of solutions having equal osmotic pressure

132

SYNTONIZE (P) [SYN-(together) + -TON-(stretch: thus to raise voice, pitch)]
  adjust or tune radio instruments or systems to the same resonant frequency

TONOMETER (C) [-TON- + -METER-(measure)]
  instrument which measures vapor pressure

*(attune, atonal, baritone, diatonic, intonation, monotony, tonicity, tonus)*

### -TOP- (place)

ISOTOPE (P) [-ISO-(equal) + -TOP-]
  one of two or more forms of an element having same atomic number and position in atomic table and similar chemical properties but different atomic weights

TOPOLOGY (M) [-TOP- + -LOGY-(science of)]
  study of the properties of geometric figures and solids that are unaffected by certain transformations (so named because it is concerned with the geometrical theory of situation without respect to size or shape)

*(topiary, topic, topical, topography, toponymy)*

### -TORT-, -TORS- (twist)

DISTORTION (P) [DIS-(apart) + -TORT-]
  aberration in a lens affecting the position of images off the axis

RETORT (C) [RE-(back) + -TORT-]
  glass vessel with a long bent tube used for heating or distillation

TORQUE (P) [TORQ < -TORT-]
  force or combination of forces that tends to produce rotation or twisting

133

TORSION (P)  [-TORS-]
   deformation of a body due to twisting

*(contort, extort, retort, torch, torment, tort, tortoise,
tortuous, torture)*

## -TRACT- (draw, pull)

ATTRACTION (P)  [AT < AD-(to) + -TRACT-]
   mutual force between bodies causing them to approach
   each other or resist separation; opposed to REPULSION
CONTRACTION (P)  [CON-(together) + -TRACT-]
   decrease in volume of a solid due to drawing together of
   atoms within molecules; opposed to EXPANSION
EXTRACTION (M)  [EX-(out) + -TRACT-]
   operation of calculating the root of a number or quantity
SUBTRACTION (M)  [SUB-(away) + -TRACT-]
   operation of deducting one quantity from another

*(abstract, detract, distract, portray, protract,
retract, tract, tractable)*

## TRANS- (across)

TRANSDUCER (P)  [TRANS- + -DUC-(to lead)]
   device supplying power in a different form to a second
   system
TRANSFORM (M, P)  [TRANS- + "form"]
   change an expression or operation into an equivalent
   form or one with similar properties; change one energy
   form into another, or change a current in potential or type
TRANSITION (C, P)  [TRANS- + -IT-(go)]
   of the three triads of elements forming Group 8 of the
   periodic table (so named because each group of three is
   the connecting link between two periods); sudden change
   in the energy state of an atom or nucleus

TRANSVERSE (M, P)   [TRANS- + -VERS-(turn)]
of the axis of a hyperbola passing through its foci; of a wave whose component particles oscillate across or perpendicular to the wave direction

*(transcend, transgression, transient, translate,
transmute, transpose)*

### -TRI- (three)

TRI- (C)   [-TRI-]
prefix indicating three atoms or radicals, as TRIBASIC, TRICHLORIDE, TRIOXIDE
TRIAD (C)   [-TRI-]
group of three elements with similar properties, or an atom or radical which has a valence of three
TRIGONOMETRY (M)   [-TRI- + -GON-(angle) +
-METR-(measure)]
branch of mathematics dealing with relationships of sides and angles of TRIANGLES
TRITIUM (P)   [-TRI-]
isotope of hydrogen which has atomic weight of three

*(triceps, tricycle, trident, trillion, trimester,
trinity, triplicate, trivial)*

### -TROCH- (wheel)

TROCHOID (M)   [-TROCH-]
plane curve generated by a point on the radius of a circle that rolls along a stationary straight line; note also EPITROCHOID (EPI-"upon"), HYPOTROCHOID (HYPO-"under")
TROCHOTRON (P)   [-TROCH- + -TRON (device)]
mass spectrograph in which paths of the electrically charged particles are trochoid

*(troche, trochee, trochlear, truck, truckle)*

# -TROP- (turn, change)

ALLOTROPY (C)  [-ALLO-(different) + -TROP-]
  property of an element existing in different forms with different properties

ENTROPY (P)  [EN-(within) + -TROP-]
  measure of amount of energy in a thermodynamic system unavailable for work; note also ISENTROPIC (-ISO-"equal"), occurring without change in entropy

ISOTROPIC (P)  [-ISO-(equal) + -TROP-]
  of materials having same physical properties in all directions; opposed to ANISOTROPIC

TROPOSPHERE (P)  [-TROP- + -SPHERE-(envelope)]
  region of the atmosphere marked by turbulence and decreasing temperature with increasing altitude

*(heliotrope, trope, trophy, tropic, tropism, tropology)*


# -TURB- (disturb)

PERTURBATION (P)  [PER-(an intensive) + -TURB-]
  deviation in motion or orbit of a heavenly body caused by force of a third body

TURBID (C)  [-TURB-]
  of a liquid which has a cloudy appearance due to a suspension of foreign particles

TURBIDIMETRY (C)  [-TURB- + -METR-(measure)]
  measurement of concentration of suspended particles in a liquid by means of a TURBIDOMETER

TURBULENCE (P)  [-TURB-]
  irregular, eddying motion in a moving fluid due to obstacles or friction

*(disturb, imperturbable, perturb, turbinate,
turbine, turbulence)*

136

## ULTRA- (beyond)

ULTRAFILTER (C)   [ULTRA- + "filter"]
   pressure filter with minute pores which holds back particles passing through ordinary filters

ULTRASONIC (P)   [ULTRA- + "sonic" < -SON-(sound)]
   of frequencies beyond that of audible sound

*(UHF, ultramarine, ultramodern, ultramontane,
ultraviolet, ultra vires)*

## UN- (not)

UNSATURATED (C)   [UN- + "saturated" < -SATUR-(full)]
   of solutions able to dissolve to a greater degree; of compounds in which some element is able to combine further with other elements

UNSTABLE (C, P)   [UN- + "stable" < -STA-(stand)]
   of compounds tending to decompose readily; of a state of equilibrium in which a body is subject to a large change from a small force

*(unawares, uncanny, uncouth, ungainly, unkempt,
unruffled, unruly)*

## -UND- (wave)

REDUNDANT (EQUATION) (M)   [RE-(back) + -UND-(wave: thus excess)]
   derived equation containing more solutions than the original equation

UNDULATION (P)   [-UND-]
   wavelike motion continuously propagated among the particles of a medium

*(abound, abundance, inundate, redound, superabundant,
undulant fever)*

137

## UNI- (one)

UNIFIED (FIELD THEORY) (P)   [UNI- + -FI(C)-
(make)]
   Einstein's theory treating electromagnetism, gravitation,
   and relativity as different aspects of the same process
UNITY (M)   [UNI-]
   the number one
UNIVALENT (C)   [UNI- + "valence" < -VAL-(strength)]
   having a valence of one
UNIVERSE (P)   [UNI- + -VERS-(turn)]
   the whole creation, the cosmos (so named from the sense
   of all existing things "turning as one")

*(unicorn, uniformity, unilateral, union, unique,
unison, unit, unite)*

## -URAN- (heavens)

TRANSURANIC (C, P)   [TRANS-(beyond) + "URANium"]
   of the radioactive elements whose atomic weights are
   greater than that of uranium
URANIUM (C, P)   ["URANus" < -URAN-]
   element (named after the planet Uranus)

*(Urania, uranic, uranography, Uranus)*

## -VAC- (empty)

EVACUATE (P)   [E-(out) + -VAC-]
   remove air from a sealed container
VACUUM (P)   [-VAC-]
   space containing air or gas at a much lower pressure than
   that of the atmosphere

*(vacancy, vacant, vacate, vacation, vacuity, vacuous)*

## -VAL- (strength)

COVALENCE (C)   [CO-(together) + "valence"
    < -VAL-]
bond of shared pairs of electrons between atoms of a compound; also number of electron pairs that can be shared between atoms of different elements

EQUIVALENT (C, M)   [-EQUI-(equal) + -VAL-]
having same valence or combining weight; having equal area but of different shape

VALENCE (C)   [-VAL-]
combining power of an element or radical, expressed by the number of hydrogen atoms one atom or radical can combine with or replace; note compounds UNIVALENT, BIVALENT

VALUE (M)   [-VAL-(be strong: thus be worth)]
quantity or amount denoted by a symbol or expression

*(avail, convalesce, countervail, evaluate, prevail, valiant, valid, valor)*

## -VAPOR- (steam)

EVAPORATION (C)   [E-(out, away) + -VAPOR-]
process of changing from liquid to gaseous state

VAPOR (C)   [-VAPOR-]
gaseous phase of a substance below its critical temperature, where increasing pressure will produce condensation; note also VAPOR PRESSURE, pressure of a confined vapor which is in equilibrium with its liquid

*(vaporing, vaporize, vaporous)*

## -VARI- (changing, differing)

INVARIANT (C, M)   [IN-(not) + -VARI-]

of a physical-chemical system possessing no degree of freedom; of a quantity that remains unchanged during a specified group of operations

VARIABLE (M)   [-VARI-]
quantity capable of having a number of different values throughout a calculation or operation

*(invariable, unvarying, variation, variegated, variety, various)*

## -VECT- (carry)

CONVECTION (P)   [CON-(together) + -VECT-]
transmission of heat in a gas or liquid by the mass movement of particles due to unequal temperatures and densities

VECTOR (M)   [-VECT-]
quantity having direction in space and magnitude, or the line representing it; distinguished from SCALAR

*(convex, invective, inveigh, vehement, vehicle)*

## -VERG- (bend)

CONVERGENT (M, P)   [CON-(together) + -VERG-]
of lines gradually coming together and meeting at a point in a finite distance, or of a series of values which gradually approaches a limit as new values are added; of a lens causing parallel light rays to come to a point focus

DIVERGENT (M, P)   [DI < DIS-(apart) + -VERG-]
of lines moving out in different directions from a point or away from each other; of a series of values which does not approach a definite limiting value; of a lens causing parallel light rays to spread out and come to no real focus

*(converge, diverge, verge)*

## -VELOP- (wrap)

ENVELOPE (M)　[EN-(in) + -VELOP-]
　　curve or surface which is tangent to each one of a series
　　of curves or surfaces

DEVELOP (M)　[DE < DIS-(apart) + -VELOP-(wrap:
　　thus expand)]
　　work out or expand a function or expression in the form
　　of a series; in geometry, unroll or flatten out a solid sur-
　　face to form a plane figure

*(developer, developmental, envelop, envelopment)*


## -VERS-, -VERT- (turn)

INVERSE (M)　[IN-(in, towards, up) + -VERS-(turn: thus
　　opposite, upside down)]
　　of a mutual relation between variables in which one in-
　　creases while the other decreases; reciprocal of a number

REVERSIBLE (C)　[RE-(back) + -VERS-]
　　of a reaction which can go in either direction depending
　　upon conditions

TRANSVERSE (M, P)　[TRANS-(across) + -VERS-]
　　of the axis of a hyperbola passing through its foci; of a
　　wave whose component particles oscillate across or per-
　　pendicular to the wave direction

VERTEX (M)　[Latin: vertex, vertic (top, turning point)
　　< VERTere (turn)]
　　point of intersection of the two sides forming an angle;
　　point of a triangle furthest from and opposite to the line
　　chosen as base

*(avert, controversy, convert, diversion, pervert,*
*subvert, version, vertigo)*

# -VITR- ( glass )

VITRIFICATION (C)   [-VITR- + -FIC-(make)]
  change of a refractory material at high temperature into
  glass or VITREOUS (glasslike) substance
VITRIOL (C)   [-VITR-]
  concentrated sulfuric acid, formerly called oil of vitriol (so
  named because made from green vitriol, iron sulfate,
  which has a glassy luster)

*(vitreous humor, vitric, vitrine, vitriolic)*


# -VOLUT-, -VOLV- ( roll )

INVOLUTE (M)   [IN-(in) + -VOLUT-(roll: thus a rolling
    or coiling inward)]
  curve traced by a fixed point on a taut string as it is un-
  wound from a fixed curve, called its EVOLUTE (E-"out")
INVOLUTION (M)   [IN-(in, upon) + -VOLUT-]
  multiplying a quantity by itself any number of times, thus
  raising a quantity to any given power; opposed to EVOLU-
  TION (E-"out"), obtaining the root
REVOLUTION (M)   [RE-(back) + -VOLUT-(roll: thus
    turn around)]
  movement of a figure about a fixed axis or center
VOLUMETRIC (C)   [Latin: volumen (roll of parchment:
    thus amount of space) < VOLVere (roll) + -METR-
    (measure)]
  of an analysis using definite amounts or volumes of
  standard solutions

*(convolution, devolution, evolve, involve, revolt,*
*voluble, voluminous)*

## -XEN- (strange)

EUXENITE (C)　[EU-(good) + -XEN-(strange: thus good to strangers, hospitable)]

    mineral (so named because it contains several rare elements)

XENON (C)　[-XEN-]

    element (so named because it was one of the rare or strange gases found in the atmosphere)

*(xenophobia, xenophobic)*

# General Scientific Suffixes

| SUFFIX | USE | MEANING | EXAMPLES |
|---|---|---|---|
| -ABLE | adj | able to (be) | commensurable, permeable |
| -AL | adj | pertaining to | numerical, peripheral |
| -AL | noun | (noun from adjective) | binomial, terminal |
| -ANCE | noun | action, state, condition | capacitance, resonance |
| -ANT | adj | having quality or action of | invariant, redundant |
| -AR | adj | pertaining to | equiangular, similar |
| -ATE | verb | (cause to) become | dehydrate, evaporate |
| -CUL | noun | diminutive | molecule, particle |
| -ENCE | noun | action, state, condition | incidence, resilience |
| -ENT | adj | having quality or action of | adjacent, translucent |
| -ENT | noun | (noun from adjective) | component, tangent |
| -ER | noun | performer of verbal action | promoter, transducer |
| -ESCE | verb | become, undergo | deliquesce, effervesce |
| -IBLE | adj | able to (be) | fusible, immiscible |
| -IC | adj | having, consisting of | concentric, periodic |
| -ION | noun | action, state, result | fusion, saturation |
| -ITY | noun | state, condition | affinity, density |
| -IVE | adj | having quality of | radioactive, siccative |
| -OID | noun | like, resembling | colloid, platinoid |
| -OID | noun | (curve) having form of | cycloid, trochoid |

| -OUS | adj | having, like | amorphous, aqueous |
| -OR | noun | performer of verbal action | accelerator, donor |
| -TRON | noun | subatomic particle device | cyclotron, synchrotron |

# Chemical Suffixes

| SUFFIX | MEANING | EXAMPLES |
|---|---|---|
| -ATE | 1) salts derived from -IC acids | nitrate, sulfate |
| | 2) verb meaning combine, treat with | chlorinate, hydrogenate |
| -IC | having a higher valence than indicated by -OUS | ferric chloride<br>nitric acid |
| -IN | neutral compounds, as fats, proteins | gelatin, stearin |
| -INE | 1) halogens | bromine, chlorine |
| | 2) alkaloids or nitrogenous bases | aniline, morphine |
| -ITE | salts derived from -OUS acids | nitrite, sulfite |
| -IUM | some radicals, elements | ammonium, magnesium |
| -OL | alcohols or phenols | glycerol, menthol |
| -ON | 1) inert gases | argon, neon |
| | 2) some non-metallic elements | boron, carbon |
| | 3) atomic particles | meson, proton |
| -OSE | carbohydrates | cellulose, dextrose |
| -OUS | having a lower valence than indicated by -IC | ferrous chloride<br>nitrous acid |